A-Z
EARTH SCIENCE

A-Z
EARTH SCIENCE

Dr. Samayal Varadarajan

CENTRUM PRESS
NEW DELHI-110002 (INDIA)

CENTRUM PRESS
H.O.: 4360/4, Ansari Road, Daryaganj,
New Delhi-110 002 (India)
Ph.: 23278000, 23261597

B.O.: No. 1015, Ist Main Road, BSK IIIrd Stage
IIIrd Phase, IIIrd Block,
Bangalore - 560 085 (India)
Tel.: 080-41723429
Visit us at: www.centrumpress.com

A-Z Earth Science

First Edition, 2009

ISBN 978-93-80106-13-7

PRINTED IN INDIA

Printed at Salasar Imaging Systems, Delhi-110035 (India)

Contents

Preface

Earth Science is an all-embracing term for the sciences related to the planet Earth. It is arguably a special case in planetary science, the Earth being the only known life-bearing planet. There are both reductionist and holistic approaches to Earth science. There are four major disciplines in earth sciences, namely *Geography, Geology, Geophysics and Geodesy*. These major disciplines use Physics, Chemistry, Biology, Chronology and Mathematics to build a quantitative understanding of the principal areas or *Spheres* of the Earth system.

After much debate based on the accumulating scientific evidence it is now generally, though still not universally, accepted that the Earth's climate is undergoing changes induced by man's activities. Global warming, due primarily to increasing concentrations of 'greenhouse gases' such as carbon dioxide in the Earth's atmosphere, has become a matter of concern well beyond the scientific community, to politicians and to the general public. The consequences of global warming are profound. Climate changes, including major effects on rainfall volumes and patterns, impact immediately on agricultural practice and thus the food supply; rising temperatures of the world's oceans and melting of polar ice will lead to flooding and coastal erosion and may have severely adverse influences on marine ecosystems. In addition to climate change, the industrial activity, which has been so vital to development of nations around the world, has other impacts of global significance, such as air pollution and declining fish stocks.

Author

Chapter 1

The Earth

The word "earth" has been used in several connexions - from that of soil or ground to that of the planet which we inhabit, but it is difficult to trace the exact historic sequence of the diverse usages. In the cosmogony of the Pythagoreans, Platonists and other philosophers, the term or its equivalent denoted an element or fundamental quality which conferred upon matter the character of earthiness; and in the subsequent development of theories as to the ultimate composition of matter by the alchemists, iatrochemists, and early phlogistonists an element of the same name was retained. In modern chemistry, the common term "earth" is applied to certain oxides: - the alkaline earths "are the oxides of calcium (lime), barium. (baryta) and strontium (strontia); the "rare earths" are the oxides of a certain class of rare metals.

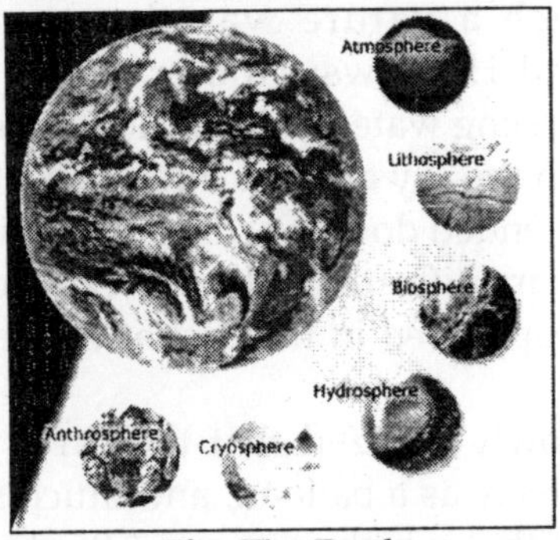

Fig. The Earth

The terrestrial globe is a member of the Solar system, the third in distance from the Sun, and the largest within the orbit of Jupiter. In the wider sense it may be regarded as composed of a gaseous atmosphere, which encircles the crust or lithosphere and surface waters or hydrosphere. The description of the surface features is a branch of Geography, and the discussions as to their origin and permanence belongs to Physiography, physiographical geology, or physical geography. The investigation of the crust belongs to geology and of rocks in particular to petrology.

In the present article we shall treat the subject matter of the Earth as a planet under the following headings:

- Figure and Size,
- Mass and Density,
- Astronomical Relations,
- Evolution and Age.

These subjects will be treated summarily,. readers being referred to the article Astronomy and to the cross-references for details.

FIGURE AND SIZE

To primitive man the Earth was a flat disk with its surface diversified by mountains, rivers and seas. In many cosmogonies this disk was encircled by waters, unmeasurable by man and extending to a junction with the sky; and the disk stood as an island rising up through the waters from the floor of the universe, or was borne as an immovable ship on the surface. Of such a nature was the cosmogony of the Babylonians and Hebrews; Homer states the same idea, naming the encircling waters and Hesiod regarded it as a disk midway between the sky and the infernal regions. The theory that the Earth extended downwards to the limit of the universe was subjected to modification when it was seen that the same sun and stars reappeared in the east after their setting in the west.

But man slowly realized that the earth was isolated in space, floating freely as a balloon, and much speculation was associated about that which supported the Earth. Tunnels. in the foundations to permit the passage of the sun and stars were suggested; the Greeks considered twelve columns to support the heavens, and in their mythology the god Atlas appears condemned to support the columns; while the Egyptians had the Earth supported by four elephants, which themselves stood on a tortoise swimming on a sea. Earthquakes were. regarded as due to a movement of these foundations; in Japan this was considered to be due to the motion of a great spider, an animal subsequently replaced by a cat-fish; in Mongolia it is a hog; in India, a mole; in some parts of South America, a whale; and among some of the North American Indians,. a giant tortoise.

The doctrine of the spherical form has been erroneously assigned to Thales; but he accepted the Semitic conception of the disk, and regarded the production of springs after earthquakes as due to the inrushing of the waters under the Earth into fissures in the surface. His pupil, Anaximander, according to Diogenes Laertius, believed it to be spherical and Anaximenes probably held a similar view. The spherical form is undoubtedly a discovery of Pythagoras, and was taught by the Pythagoreans and by the Eleatic Parmenides.

The expositor of greatest moment was Aristotle; his arguments arė those which we employ to-day:- the ship gradually disappearing from hull to mast as it recedes from the harbour to the horizon; the circular shadow cast by the Earth on the Moon during an eclipse, and the alteration in the appearance of the heavens as one passes from point to point on the Earth's surface.' He records attempts made to determine the circumference; but the first scientific investigation in this

1 Aristotle regarded the Earth as having an upper inhabited half and a lower uninhabited one, and the air on the lower half as tending to flow upwards through the Earth. The obstruction of this passage brought about an accumulation of air within the Earth, and the increased pressure may occasion oscillations of the surface, which may be so intense as to cause earthquakes.

The historical development of the methods for determining the figure of the Earth (by which we mean a theoretical surface in part indicated by the ocean at rest, and in other parts by the level to which water freely communicating with the oceans by canals traversing the land masses would rise) and the mathematical investigation of this problem are treated in the articles Figure of the Earth, and Geodesy; here the results are summarized. Sir Isaac Newton deduced from the mechanical consideration of the figure of equilibrium of a mass of rotating fluid, the form of an oblate spheroid, the ellipticity of a meridian section being 1/231, and the axes in the ratio 230:231.

Geodetic measurements by the Cassinis and other French astronomers pointed to a prolate form, but the Newtonian figure was proved to be correct by the measurement of meridional arcs in Peru and Lapland by the expeditions organized by the French Academy of Sciences. The position of the longer axis is somewhat uncertain; it is certainly in Africa, Clarke placing it in longitude 8° 15' W., and Schubert in longitude 41° 4' E.; W. J. Sollas, arguing from terrestrial symmetry, has chosen the position lat. 6° N., long. 28° E., *i.e.* between Clarke's and Schubert's positions. For the lengths of the axes and the ellipticity of the Earth.

Mass and Density

The earliest scientific investigation on the density and mass of the Earth (the problem is really single if the volume of the Earth be known) was made by Newton, who, mainly from astronomical considerations, suggested the limiting densities 5 and 6; it is remarkable that this prophetic guess should be realized, the mean value from subsequent researches being about 5.2, which gives for the mass the value 6 X io 2 ‘ tons. The density of the Earth has been determined by several experimenters within recent years by methods.

Astronomical Relations

The grandest achievements of astronomical science are undoubtedly to be associated with the elucidation of the complex motion of our planet. The notion that the Earth was fixed and immovable at the centre of an immeasurable universe long possessed the minds of men; and we find the illustrious Ptolemy accepting this view in the 2nd century A.D.. and rejecting the notion of a rotating Earth - a theory which had been proposed as early as the 5th century B.C. by Philolaus on philosophical grounds, and in the 3rd century B.C. by the astronomer Aristarchus of Samos. He argued that if the Earth rotated then points at the equator had the enormous velocity of about 1000 m. per hour, and as a consequence there should be terrific gales from the east; the fact that there were no such gales invalidated, in his opinion, the theory.

The Ptolemaic theory was unchallenged until 1543, in which year the *De Revolutionibus orbium Celestium* of Copernicus was published. In this work it was shown that the common astronomical phenomena could be more simply explained by regarding the Earth as annually revolving about a fixed Sun, and daily rotating about itself. A clean sweep was made of the geocentric epicyclic motions of the planets which Ptolemy's theory demanded, and in place there was substituted a procession of planets about the Sun at different distances.The Earth has two principal motions - revolution about the Sun, rotation about its axis; there are in addition a number of secular motions.

REVOLUTION

The Earth revolves about the Sun in an elliptical orbit having the Sun at one focus. The plane of the orbit is termed the ecliptic; it is inclined to the Earth's equator at an angle termed the obliquity, and the points of intersection of the equator and ecliptic are termed the equinoctial points. The major axis of the ellipse is the line of apsides; when the Earth is nearest the Sun it is said to be in perihelion, when farthest it is in aphelion. The mean distance of the Earth from the Sun is a most important astronomical constant, since it is the unit of linear measurement; its value is about 93,000,000 m., and the difference between the perihelion and aphelion distances is about 3,000,000 m.

The eccentricity of the orbit is 016751. A tabular comparison of the orbital constants of the Earth and the other planets is given in the article Planet. The period of revolution with regard to the Sun, or, in other words, the time taken by the Sun apparently to pass from one equinox to the same equinox, is the tropical or equinoctial year; its length is 365 d.

5 hrs. 48 m. 46 secs. It is about 20 minutes shorter than the true or sidereal year, which is the time taken for the Sun apparently to travel from one star to it again. The difference in these two years is due to the secular variation termed precession. A third year is named the *anomalistic year,* which is the time occupied in the passage from perihelion to perihelion; it is a little longer than the sidereal.

ROTATION

The Earth rotates about an axis terminating at the north and south geographical poles, and perpendicular to the equator; the period of rotation is termed the day, of which several kinds are distinguished according to the body or point of reference. The rotation is performed from west to east; this daily rotation occasions the *diurnal* motion of the celestial sphere, the rising of the Sun and stars in the east and their setting in the west, and also the phenomena of day and night. The inclination of the axis to the ecliptic brings about the presentation of places in different latitudes to the more direct rays of the sun; this is revealed in the variation in the length of daylight with the time of the year, and the phenomena of seasons.

Although the rotation of the Earth was an accepted fact soon after its suggestion by Copernicus, an experimental proof was wanting until 1851, when Foucault performed his celebrated pendulum experiment at the Pantheon, Paris. A pendulum about 200 ft. long, composed of a flexible wire carrying a heavy iron bob, was suspended so as to be free to oscillate in any direction.

The bob was provided with a style which passed over a table strewn with fine sand, so that the style traced the direction in which the bob was swinging. It was found that the oscillating pendulum never retraced its path, but at each swing it was apparently deviated to the right, and moreover the deviations in equal times were themselves equal. This means that the floor of the Pantheon was moving, and therefore the Earth was rotating. If the pendulum were swung in the southern hemisphere, the deviation would be to the left;

if at the equator it would not deviate, while at the poles the plane of oscillation would traverse a complete circle in 24 hours. The rotation of the Earth appears to be perfectly uniform, comparisons of the times of transits, eclipses, etc, point to a variation of less than yhth of a second since the time of Ptolemy. Theoretical investigations on the phenomena of tidal friction point, however, to a retardation, which may to some extent be diminished by the accelerations occasioned by the shrinkage of the globe, and some other factors difficult to evaluate.

Precession

The axis of the earth does not preserve an invariable direction in space, but in a certain time it describes a cone, in much the same manner as the axis of a top spinning out of the vertical. The equator, which preserves approximately the same inclination to the ecliptic must move so that its intersections with the ecliptic, or equinoctial points, pass in a retrograde direction, *i.e.* opposite to that of the Earth. This motion is termed the precession of the equinoxes, and was observed by Hipparchus in the 2nd century B.C.; Ptolemy corrected the catalogue of Hipparchus for precession by adding 2° 40' to the longitudes, the latitudes being unaltered by this motion, which at the present time is 50.26" annually, the complete circuit being made in about 26,000 years.

Owing to precession the signs of the zodiac are traversing paths through the constellations, or, in other words, the constellations are continually shifting with regard to the equinoctial points; at one time the vernal equinox Aries was in the constellations of that name; it is now in Pisces, and will then pass into Aquarius. The pole star, *i.e.* the star towards which the Earth's axis points, is also shifting owing to precession; in about 2700 B.C. the Chinese observed a Draconis as the pole star (at present a Ursae minoris occupies this position and will do so until 3500); in 13600 Vega (a Lyrae) the brightest star in the Northern hemisphere, will be nearest.

Precession is the result of the Sun and the Moon's attraction on the Earth not being a single force through its centre of gravity. If the Earth were a homogeneous sphere the attractions would act through the centre, and such forces would have no effect upon the rotation about the centre of gravity, but the Earth being spheroidal the equatorial band which stands up as it were beyond the surface of a sphere is more strongly attracted, with the result that the axis undergoes a tilting. The precession due to the Sun is termed the *solar precession* and that due to the Moon the *lunar precession;* the joint effect (two-thirds of which is due to the Moon) is the *luni-solar* precession. Solar precession is greatest at the solstices and zero at the equinoxes; the part of luni-solar precession due to the Moon varies with the position of the Moon in its orbit. The obliquity is unchanged by precession.

Nutation

In treating precession we have stated that the axis of the Earth traces a cone, and it follows that the pole describes a circle on the celestial sphere, about the pole of the ecliptic. This is not quite true. Irregularities in the attracting forces which occasion precession also cause a slight oscillation backwards and forwards over the mean precessional path of the pole, the pole tracing a wavy line or nodding. Both the Sun and Moon contribute to this effect. Solar nutation depends upon the position of the Sun on the ecliptic; its period is therefore I year, and in extent it is only I.2 lunar nutation depends upon the position of the Moon's nodes; its period is therefore about 18.6 years, the time of revolution of the nodes, and its extent is 9. 2". There is also given to the obliquity a small oscillation to and fro. Nutation is one of the great discoveries of James Bradley.

Planetary Precession

So far we have regarded the ecliptic as absolutely fixed, and treated precession as a real motion of the equator. The ecliptic, however, is itself subject to a motion, due to the attractions of the planets on the Earth. This effect also displaces the equinoctial points. Its annual value is 0.13". The term General Precession in longitude is given to the displacement of the intersection of the equator with the apparent ecliptic on the latter. The standard value is 50.2453", which prevailed in 1850, and the value at 1850+t, *i.e.* the constant of precession, is 5 0.2 453" + 0.0002225" *t*. This value is also liable to a very small change. The nutation of the obliquity at time 1850 + t is given by the formula 23° 27' 32. o" -0.47" I. Complete expressions for these functions are given in Newcomb's *Spherical Astronomy* and in the *Nautical Almanac*. The variation of the *line of apsides* is the name given to the motion of the major axis of the Earth's orbit along the ecliptic. It is due to the general influence of the planets, and the revolution is effected in 21,000 years.

The variation of the eccentricity denotes an oscillation of the form of the Earth's orbit between a circle and ellipse. This followed the mathematical researches of Lagrange and Leverrier. It was suggested by Sir John Herschel in 1830 that this variation might occasion great climatic changes, and James Croll developed the theory as affording a solution of the glacial periods in geology.

Variation of Latitude

Another secular motion of the Earth is due to the fact that the axis of rotation is not rigidly fixed within it, but its polar extremities wander in a circle of about 50 ft. diameter. This oscillation brings about a variability in terrestrial latitudes, hence the name. Euler showed mathematically that such an oscillation existed, and, making certain assumptions as to the rigidity of the Earth, deduced that its period was 305 days; S. C. Chandler, from 1890 onwards, deduced from observations of the stars a period of 428 days;. and Simon Newcomb explained the deviation of these periods by pointing out that Euler's assumption of a perfectly rigid Earth is not in accordance with fact.

EVOLUTION AND AGE

In its earliest history the mass now consolidated as the Earth and Moon was part of a vast nebulous aggregate, which in the course of time formed a central nucleus - our Sun - which shed its outer layers in such a manner as to form the solar system. The moon may have been formed from the Earth in a similar manner, but the theory of tidal friction suggests the elongation of the Earth along an equatorial axis to form a pear-shaped figure, and that in the course of time the protuberance shot off to form the Moon. The age of the Earth has been investigated from several directions, as have also associated questions related to climatic changes, internal temperature, orientation of the land and water (permanence of oceans and continents), etc. These problems are treated in the articles Geology and Geography.

Figure of the Earth

The determination of the figure of the earth is a problem of the highest importance in astronomy, inasmuch as the diameter of the earth is the unit to which all celestial distances must be referred.

Scientists have been able to reconstruct detailed information about the planet's past. Earth and the other planets in the Solar System formed 4.54 billion years ago out of the solar nebula, a disk-shaped mass of dust and gas left over from the formation of the Sun. Initially molten, the outer layer of the planet Earth cooled to form a solid crust when water began accumulating in the atmosphere. The Moon formed soon afterwards, possibly as the result of a Mars-sized object (sometimes called Theia) with about 10% of the Earth's mass impacting the Earth in a glancing blow. Some of this object's mass would have merged with the Earth and a portion would have been ejected into space, but enough material would have been sent into orbit to form the Moon.

Outgassing and volcanic activity produced the primordial atmosphere. Condensing water vapour, augmented by ice and liquid water delivered by asteroids and the larger proto-planets, comets, and trans-Neptunian objects produced the oceans. The highly energetic chemistry is believed to have produced a self-replicating molecule around 4 billion years ago, and half a billion years later, the last common ancestor of all life existed.

The development of photosynthesis allowed the Sun's energy to be harvested directly by life forms; the resultant oxygen accumulated in the atmosphere and resulted in a layer of ozone (a form of molecular oxygen [O_3]) in the upper atmosphere.

The incorporation of smaller cells within larger ones resulted in the development of complex cells called eukaryotes. True multicellular organisms formed as cells within colonies became increasingly specialized. Aided by the absorption of harmful ultraviolet radiation by the ozone layer, life colonized the surface of Earth.

Beginning with almost no dry land, the total amount of surface lying above the oceans has steadily increased. During the past two billion years, for example, the total size of the continents has doubled. As the surface continually reshaped itself, over hundreds of millions of years, continents formed and broke up. The continents migrated across the surface, occasionally combining to form a supercontinent. Roughly 750 million years ago, the earliest known supercontinent, Rodinia, began to break apart. The continents later recombined to form Pannotia, 600–540 mya, then finally Pangaea, which broke apart 180 mya.

Since the 1960s, it has been hypothesized that severe glacial action between 750 and 580 mya, during the Neoproterozoic, covered much of the planet in a sheet of ice. This hypothesis has been termed "Snowball Earth", and is of particular interest because it preceded the Cambrian explosion, when multicellular life forms began to proliferate.

Following the Cambrian explosion, about 535 mya, there have been five mass extinctions. The last extinction event occurred 65 mya, when a meteorite collision probably triggered the extinction of the (non-avian) dinosaurs and other large reptiles, but spared small animals such as mammals, which then resembled shrews.

Over the past 65 million years, mammalian life has diversified, and several mya, an African ape-like animal gained the ability to stand upright. This enabled tool use and encouraged communication that provided the nutrition and stimulation needed for a larger brain. The development of agriculture, and then civilization, allowed humans to influence the Earth in a short time span as no other life form had, affecting both the nature and quantity of other life forms.

The present pattern of ice ages began about 40 mya, then intensified during the Pleistocene about 3 mya. The polar regions have since undergone repeated cycles of glaciation and thaw, repeating every 40–100,000 years. The last ice age ended 10,000 years ago.

COMPOSITION AND STRUCTURE

Earth is a terrestrial planet, meaning that it is a rocky body, rather than a gas giant like Jupiter. It is the largest of the four solar terrestrial planets, both in terms of size and mass. Of these four planets, Earth also has the highest density, the highest surface gravity and the strongest magnetic field.

SHAPE

Chemical Composition

The mass of the Earth is approximately 5.98×10 kg. It is composed mostly of iron (32.1%), oxygen (30.1%), silicon (15.1%), magnesium (13.9%), sulfur (2.9%), nickel (1.8%), calcium (1.5%), and aluminium (1.4%); with the remaining 1.2% consisting of trace amounts of other elements. Due to mass segregation, the core region is believed to be primarily composed of iron (88.8%), with smaller amounts of nickel (5.8%), sulfur (4.5%), and less than 1% trace elements.

The geochemist F. W. Clarke calculated that a little more than 47% of the Earth's crust consists of oxygen. The more common rock constituents of the Earth's crust are nearly all oxides; chlorine, sulfur and fluorine are the only important exceptions to this and their total amount in any rock is usually much less than 1%.

The principal oxides are silica, alumina, iron oxides, lime, magnesia, potash and soda. The silica functions principally as an acid, forming silicates, and all the commonest minerals of igneous rocks are of this nature. From a computation based on 1,672 analyses of all kinds of rocks, Clarke deduced that 99.22% were composed of 11 oxides. All the other constituents occur only in very small quantities.

Internal Structure

The interior of the Earth, like that of the other terrestrial planets, is chemically divided into layers. The Earth has an outer silicate solid crust, a highly viscous mantle, a liquid outer core that is much less viscous than the mantle, and a solid inner core. The crust is separated from the mantle by the Mohorovièiæ discontinuity, and the thickness of the crust varies: averaging 6 km under the oceans and 30–50 km on the continents.

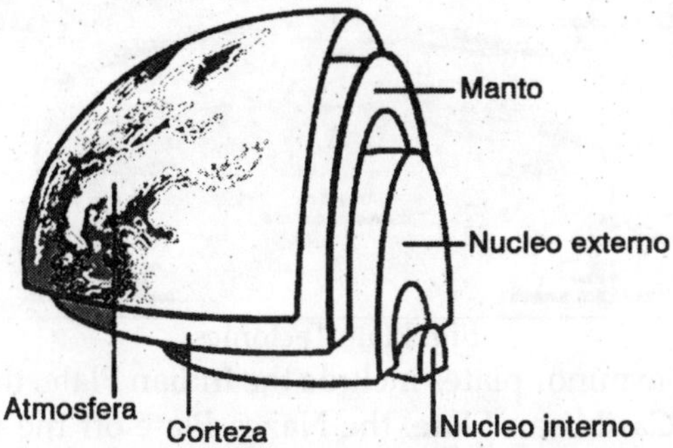

Fig. Internal Structure of the Earth

The internal heat of the planet is probably produced by the radioactive decay of potassium-40, uranium-238 and thorium-232 isotopes. All three have half-life decay periods of more than a billion years. At the centre of the planet, the temperature may be up to 7,000 K and the pressure could reach 360 GPa. A portion of the core's thermal energy is transported toward the crust by Mantle plumes; a form of convection consisting of upwellings of higher-temperature rock. These plumes can produce hotspots and flood basalts.

Tectonic Plates

According to plate tectonics theory, the outermost part of the Earth's interior is made up of two layers: the lithosphere, comprising the crust, and the solidified uppermost part of the mantle. Below the lithosphere lies the asthenosphere, which forms the inner part of the upper mantle. The asthenosphere behaves like a superheated material that is in a semi-fluidic, plastic-like state.The lithosphere essentially *floats* on the asthenosphere and is broken up into what are called tectonic plates. These plates are rigid segments that move in relation to one another at one of three types of plate boundaries: convergent, divergent and transform. The last occurs where two plates move laterally relative to each other, creating a strike-slip fault. Earthquakes, volcanic activity, mountain-building, and oceanic trench formation can occur along these plate boundaries.

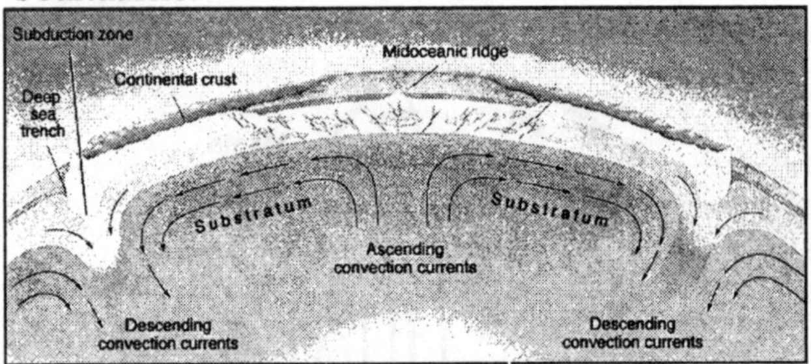

Fig. Plate Tectonics

Notable minor plates include the Indian Plate, the Arabian Plate, the Caribbean Plate, the Nazca Plate off the west coast of South America and the Scotia Plate in the southern Atlantic Ocean. The Australian Plate actually fused with Indian Plate between 50 and 55 million years ago. The fastest-moving plates are the oceanic plates, with the Cocos Plate advancing at a rate of 75 mm/yr and the Pacific Plate moving 52–69 mm/yr. At the other extreme, the slowest-moving plate is the Eurasian Plate, progressing at a typical rate of about 21 mm/yr.

Surface

The Earth's terrain varies greatly from place to place. About 70.8% of the surface is covered by water, with much of the continental shelf below sea level. The submerged surface has mountainous features, including a globe-spanning mid-ocean ridge system, as well as undersea volcanoes, oceanic trenches, submarine canyons, oceanic plateaus and abyssal plains. The remaining 29.2% not covered by water consists of mountains, deserts, plains, plateaus, and other geomorphologies.The planetary surface undergoes reshaping over geological time periods due to the effects of tectonics and erosion. The surface features built up or deformed through plate tectonics are subject to steady weathering from precipitation, thermal cycles, and chemical effects. Glaciation, coastal erosion, the build-up of coral reefs, and large meteorite impacts also act to reshape the landscape.

As the tectonic plates migrate across the planet, the ocean floor is subducted under the leading edges. At the same time, upwellings of mantle material create a divergent boundary along mid-ocean ridges. The combination of these processes continually recycles the oceanic crustal material. Most of the ocean floor is less than 100 million years in age. The oldest oceanic crust is located in the Western Pacific, and has an estimated age of about 200 million years. By comparison, the oldest fossils found on land have an age of about 3 billion years.

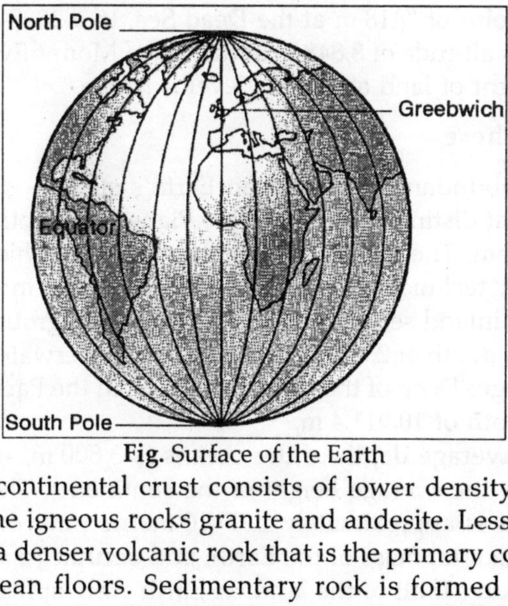

Fig. Surface of the Earth

The continental crust consists of lower density material such as the igneous rocks granite and andesite. Less common is basalt, a denser volcanic rock that is the primary constituent of the ocean floors. Sedimentary rock is formed from the accumulation of sediment that becomes compacted together.

Nearly 75% of the continental surfaces are covered by sedimentary rocks, although they form only about 5% of the crust. The third form of rock material found on Earth is metamorphic rock, which is created from the transformation of pre-existing rock types through high pressures, high temperatures, or both. The most abundant silicate minerals on the Earth's surface include quartz, the feldspars, amphibole, mica, pyroxene and olivine.

Common carbonate minerals include calcite (found in limestone), aragonite and dolomite. The pedosphere is the outermost layer of the Earth that is composed of soil and subject to soil formation processes. It exists at the interface of the lithosphere, atmosphere, hydrosphere and biosphere. Currently the total arable land is 13.31% of the land surface, with only 4.71% supporting permanent crops. Close to 40% of the Earth's land surface is presently used for cropland and pasture, or an estimated 1.3×10 km^2 of cropland and 3.4×10 km^2 of pastureland.

The elevation of the land surface of the Earth varies from the low point of "418 m at the Dead Sea, to a 2005-estimated maximum altitude of 8,848 m at the top of Mount Everest. The mean height of land above sea level is 840 m.

Hydrosphere

The abundance of water on Earth's surface is a unique feature that distinguishes the "Blue Planet" from others in the solar system. The Earth's hydrosphere consists chiefly of the oceans, but technically includes all water surfaces in the world, including inland seas, lakes, rivers, and underground waters down to a depth of 2,000 m. The deepest underwater location is Challenger Deep of the Mariana Trench in the Pacific Ocean with a depth of 10,911.4 m.

The average depth of the oceans is 3,800 m, more than four times the average height of the continents. The mass of the oceans is approximately 1.35×10 metric tons, or about 1/4400 of the total mass of the Earth, and occupies a volume of 1.386×10 km^3. If all of the land on Earth were spread evenly, water would rise to an altitude of more than 2.7 km. About

97.5% of the water is saline, while the remaining 2.5% is fresh water. The majority of the fresh water, about 68.7%, is currently in the form of ice.

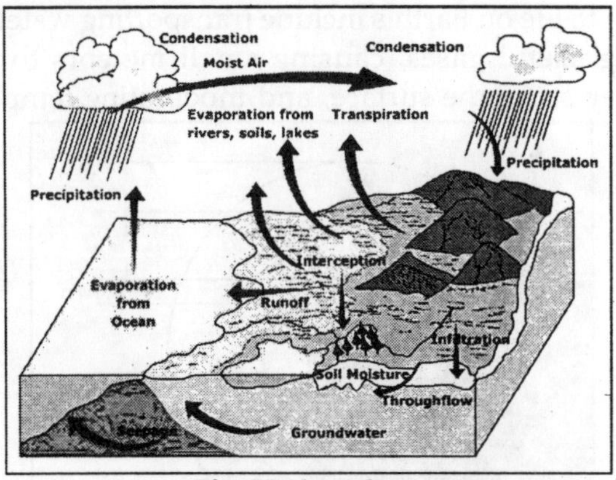

Fig. Hydrosphere

About 3.5% of the total mass of the oceans consists of salt. Most of this salt was released from volcanic activity or extracted from cool, igneous rocks. The oceans are also a reservoir of dissolved atmospheric gases, which are essential for the survival of many aquatic life forms. Sea water has an important influence on the world's climate, with the oceans acting as a large heat reservoir. Shifts in the oceanic temperature distribution can cause significant weather shifts, such as the El Niño-Southern Oscillation.

Atmosphere

The atmospheric pressure on the surface of the Earth averages 101.325 kPa, with a scale height of about 8.5 km. It is 78% nitrogen and 21% oxygen, with trace amounts of water vapour, carbon dioxide and other gaseous molecules. The height of the troposphere varies with latitude, ranging between 8 km at the poles to 17 km at the equator, with some variation due to weather and seasonal factors.

Earth's biosphere has significantly altered its atmosphere. Oxygenic photosynthesis evolved 2.7 billion years ago, forming the primarily nitrogen-oxygen atmosphere that exists today. This change enabled the proliferation of aerobic organisms as well as the formation of the ozone layer which, together with Earth's magnetic field, blocks ultraviolet solar radiation, permitting life on land. Other atmospheric functions important to life on Earth's include transporting water vapour, providing useful gases, causing small meteors to burn up before they strike the surface, and moderating temperature.

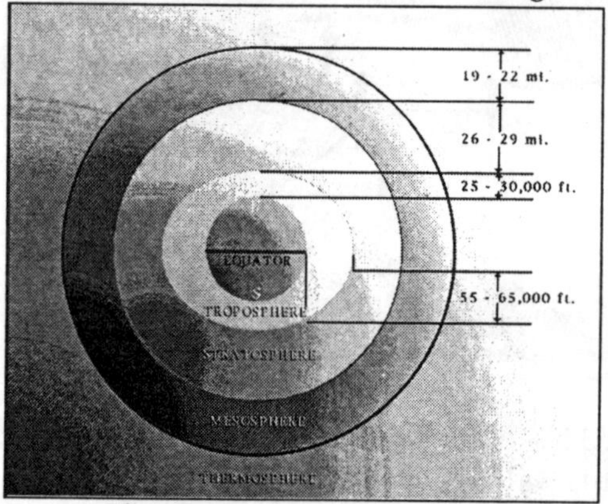

Fig. Atmosphere of the Earth

This last phenomenon is known as the greenhouse effect: trace molecules within the atmosphere serve to capture thermal energy emitted from the ground, thereby raising the average temperature. Carbon dioxide, water vapour, methane and ozone are the primary greenhouse gases in the Earth's atmosphere. Without this heat-retention effect, the average surface temperature would be "18 °C and life would likely not exist.

WEATHER AND CLIMATE

The Earth's atmosphere has no definite boundary, slowly becoming thinner and fading into outer space. Three-quarters of the atmosphere's mass is contained within the first 11 km of the planet's surface. This lowest layer is called the troposphere. Energy from the Sun heats this layer, and the surface below, causing expansion of the air. This lower density air then rises, and is replaced by cooler, higher density air. The result is atmospheric circulation that drives the weather and climate through redistribution of heat energy. The primary atmospheric circulation bands consist of the trade winds in the equatorial region below 30° latitude and the westerlies in the mid-latitudes between 30° and 60°. Ocean currents are also important factors in determining climate, particularly the thermohaline circulation that distributes heat energy from the equatorial oceans to the polar regions.

Water vapour generated through surface evaporation is transported by circulatory patterns in the atmosphere. When atmospheric conditions permit an uplift of warm, humid air, this water condenses and settles to the surface as precipitation. Most of the water is then transported back to lower elevations by river systems, usually returning to the oceans or being deposited into lakes.

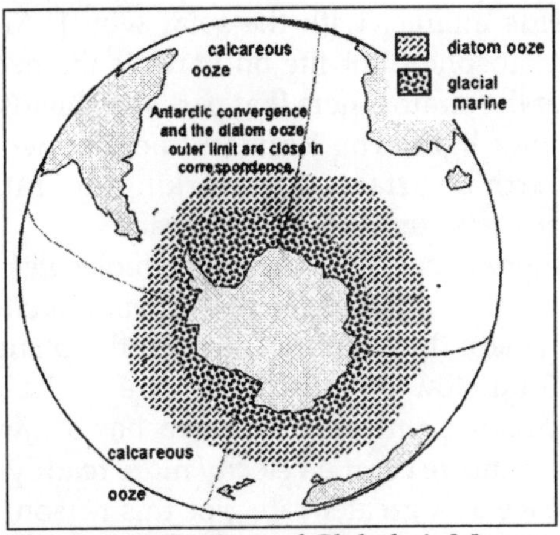

Fig. Source Regions of Global air Masses.

This water cycle is a vital mechanism for supporting life on land, and is a primary factor in the erosion of surface features over geological periods. Precipitation patterns vary widely, ranging from several meters of water per year to less than a millimeter. Atmospheric circulation, topological features and temperature differences determine the average precipitation that falls in each region.

The Earth can be sub-divided into specific latitudinal belts of approximately homogeneous climate. Ranging from the equator to the polar regions, these are the tropical (or equatorial), subtropical, temperate and polar climates. Climate can also be classified based on the temperature and precipitation, with the climate regions characterized by fairly uniform air masses. The commonly-used Köppen climate classification system has five broad groups (humid tropics, arid, humid middle latitudes, continental and cold polar), which are further divided into more specific subtypes.

Upper Atmosphere

Above the troposphere, the atmosphere is usually divided into the stratosphere, mesosphere, and thermosphere. Each of these layers has a different lapse rate, defining the rate of change in temperature with height. Beyond these, the exosphere thins out into the magnetosphere (where the Earth's magnetic fields interact with the solar wind). An important part of the atmosphere for life on Earth is the ozone layer, a component of the stratosphere that partially shields the surface from ultraviolet light. The Kármán line, defined as 100 km above the Earth's surface, is a working definition for the boundary between atmosphere and space.

Due to thermal energy, some of the molecules at the outer edge of the Earth's atmosphere have their velocity increased to the point where they can escape from the planet's gravity. This results in a slow but steady leakage of the atmosphere into space. Because unfixed hydrogen has a low molecular weight, it can achieve escape velocity more readily and it leaks into outer space at a greater rate. For this reason, the Earth's current environment is oxidizing, rather than reducing, with consequences for the chemical nature of life which developed on the planet. The oxygen-rich atmosphere also preserves much of the surviving hydrogen by locking it up in water molecules.

MAGNETIC FIELD

The Earth's magnetic field is shaped roughly as a magnetic dipole, with the poles currently located proximate to the planet's geographic poles. According to dynamo theory, the field is generated within the molten outer core region where heat creates convection motions of conducting materials, generating electric currents. These in turn produce the Earth's magnetic field. The convection movements in the core are chaotic in nature, and periodically change alignment. This results in field reversals at irregular intervals averaging a few times every million years. The most recent reversal occurred approximately 700,000 years ago.

The field forms the magnetosphere, which deflects particles in the solar wind. The sunward edge of the bow shock is located at about 13 times the radius of the Earth.

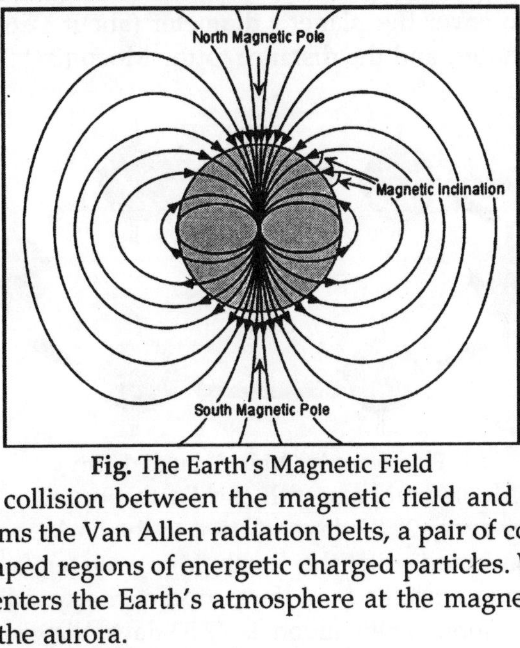

Fig. The Earth's Magnetic Field

The collision between the magnetic field and the solar wind forms the Van Allen radiation belts, a pair of concentric, torus-shaped regions of energetic charged particles. When the plasma enters the Earth's atmosphere at the magnetic poles, it forms the aurora.

ORBIT AND ROTATION

Relative to the background stars, it takes the Earth, on average, 23 hours, 56 minutes and 4.091 seconds (one sidereal day) to rotate around the axis that connects the north and the south poles from west to east. From Earth, the main apparent motion of celestial bodies in the sky (except that of meteors within the atmosphere and low-orbiting satellites) is to the west at a rate of 15°/h = 15'/min. This is equivalent to an apparent diameter of the Sun or Moon every two minutes.

Earth orbits the Sun at an average distance of about 150 million kilometers every 365.2564 mean solar days (1 sidereal year). From Earth, this gives an apparent movement of the Sun with respect to the stars at a rate of about 1°/day (or a Sun or Moon diameter every 12 hours) eastward. Because of this motion, on average it takes 24 hours—a solar day—for Earth to complete a full rotation about its axis so that the Sun returns to the meridian. The orbital speed of the Earth averages about 30 km/s (108,000 km/h), which is fast enough to cover the planet's diameter (about 12,600 km) in seven minutes, and the distance to the Moon (384,000 km) in four hours.

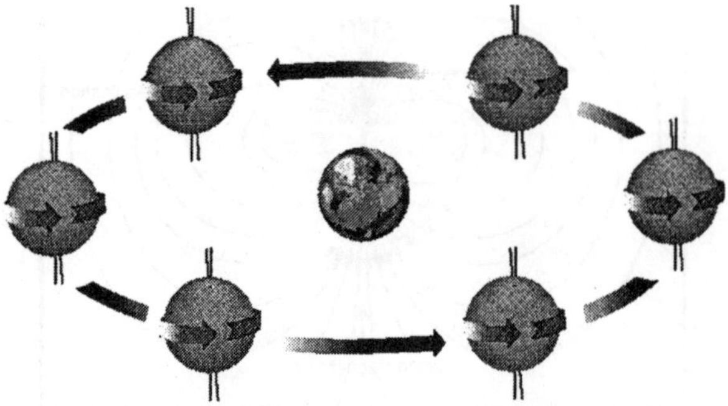

Fig. Moon's Orbit Around Earth

The Moon revolves with the Earth around a common barycenter every 27.32 days relative to the background stars. When combined with the Earth–Moon system's common revolution around the Sun, the period of the synodic month, from new moon to new moon, is 29.53 days. Viewed from the celestial north pole, the motion of Earth, the Moon and their axial rotations are all counter-clockwise. The orbital and axial planes are not precisely aligned: Earth's axis is tilted some 23.5 degrees from the perpendicular to the Earth–Sun plane (which causes the seasons); and the Earth–Moon plane is tilted about 5 degrees against the Earth-Sun plane (without this tilt, there would be an eclipse every two weeks, alternating between lunar eclipses and solar eclipses).

Because of the axial tilt of the Earth, the position of the Sun in the sky varies over the course of the year. For an observer at a northern latitude, when the northern pole is tilted toward the Sun the day lasts longer and the Sun climbs higher in the sky. This results in warmer average temperatures from the increase in solar radiation reaching the surface. When the northern pole is tilted away from the Sun, the reverse is true and the climate is generally cooler. Above the arctic circle, an extreme case is reached where there is no daylight at all for part of the year. This variation in the climate (because of the direction of the Earth's axial tilt) results in the seasons. By astronomical convention, the four seasons are determined by the solstices—the point in the orbit of maximum axial tilt toward or away from the Sun—and the equinoxes, when the direction of the tilt and the direction to the Sun are perpendicular. Winter solstice occurs on about December 21, summer solstice is near June 21, spring equinox is around March 20 and autumnal equinox is about September 23. The axial tilt in the southern hemisphere is exactly the opposite of the direction in the northern hemisphere. Thus the seasonal effects in the south are reversed.The angle of the Earth's tilt is relatively stable over long periods of time.

However, the tilt does undergo a slight, irregular motion (known as nutation) with a main period of 18.6 years. The orientation (rather than the angle) of the Earth's axis also changes over time, precessing around in a complete circle over each 25,800 year cycle; this precession is the reason for the difference between a sidereal year and a tropical year. Both of these motions are caused by the varying attraction of the Sun and Moon on the Earth's equatorial bulge. From the perspective of the Earth, the poles also migrate a few meters across the surface. This polar motion has multiple, cyclical components, which collectively are termed quasiperiodic motion. In addition to an annual component to this motion, there is a 14-month cycle called the Chandler wobble. The rotational velocity of the Earth also varies in a phenomenon known as length of day variation. The changing Earth-Sun distance results in an increase of about 6.9% in solar energy reaching the Earth at perihelion relative to aphelion. Since the southern hemisphere is tilted toward the Sun at about the same time that the Earth reaches the closest approach to the Sun, the southern hemisphere receives slightly more energy from the Sun than does the northern over the course of a year.

However, this effect is much less significant than the total energy change due to the axial tilt, and most of the excess energy is absorbed by the higher proportion of water in the southern hemisphere.The Hill sphere (gravitational sphere of influence) of the Earth is about 1.5 Gm (or 1,500,000 kilometers) in radius. This is maximum distance at which the Earth's gravitational influence is stronger than the more distant Sun and planets. Objects must orbit the Earth within this radius, or they can become unbound by the gravitational perturbation of the Sun.Earth, along with the Solar System, is situated in the Milky Way galaxy, orbiting about 28,000 light years from the centre of the galaxy, and about 20 light years above the galaxy's equatorial plane in the Orion spiral arm.

MOON

The Moon is a relatively large, terrestrial, planet-like satellite, with a diameter about one-quarter of the Earth's. It is the largest moon in the solar system relative to the size of its planet. (Charon is larger relative to the dwarf planet Pluto.) The natural satellites orbiting other planets are called "moons" after Earth's Moon.

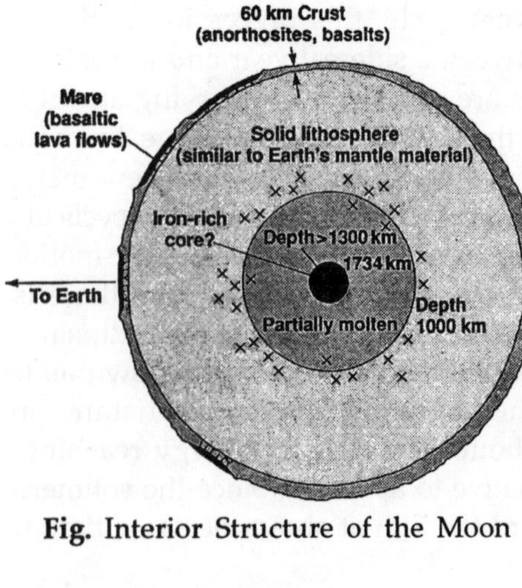

Fig. Interior Structure of the Moon

The gravitational attraction between the Earth and Moon causes tides on Earth. The same effect on the Moon has led to its tidal locking: its rotation period is the same as the time it takes to orbit the Earth. As a result, it always presents the same face to the planet. As the Moon orbits Earth, different parts of its face are illuminated by the Sun, leading to the lunar phases. The dark part of the face is separated from the light part by the solar terminator.Because of their tidal interaction, the Moon recedes from Earth at the rate of approximately 38 mm a year. Over millions of years, these tiny modifications—and the lengthening of Earth's day by about 23 μs a year—add up to significant changes. During the Devonian period, for example, (approximately 410 million years ago) there were 400 days in a year, with each day lasting 21.8 hours.

The Moon may have dramatically affected the development of life by moderating the planet's climate. Paleontological evidence and computer simulations show that Earth's axial tilt is stabilized by tidal interactions with the Moon. Some theorists believe that without this stabilization against the torques applied by the Sun and planets to the Earth's equatorial bulge, the rotational axis might be chaotically unstable, as it appears to be for Mars. If Earth's axis of rotation were to approach the plane of the ecliptic, extremely severe weather could result from the resulting extreme seasonal differences. One pole would be pointed directly toward the Sun during *summer* and directly away during *winter*. Planetary scientists who have studied the effect claim that this might kill all large animal and higher plant life. However, this is a controversial subject, and further studies of Mars—which has a similar rotation period and axial tilt as Earth, but not its large Moon or liquid core—may settle the matter.

Viewed from Earth, the Moon is just far enough away to have very nearly the same apparent-sized disk as the Sun. The angular size (or solid angle) of these two bodies match because, although the Sun's diameter is about 400 times as large as the Moon's, it is also 400 times more distant. This allows total and annular eclipses to occur on Earth.The most widely accepted theory of the Moon's origin, the giant impact theory, states that it formed from the collision of a Mars-size protoplanet called Theia with the early Earth. This hypothesis explains (among other things) the Moon's relative lack of iron and volatile elements, and the fact that its composition is nearly identical to that of the Earth's crust. Earth has at least two co-orbital asteroids, 3753 Cruithne and 2002 AA_{29}.

HABITABILITY

A planet that can sustain life is termed habitable, even if life did not originate there. The Earth provides the (currently understood) requisite conditions of liquid water, an environment where complex organic molecules can assemble, and sufficient energy to sustain metabolism. The distance of the Earth from the Sun, as well as its orbital eccentricity, rate of rotation, axial tilt, geological history, sustaining atmosphere and protective magnetic field all contribute to the conditions necessary to originate and sustain life on this planet.

Biosphere

The planet's life forms are sometimes said to form a "biosphere". This biosphere is generally believed to have begun evolving about 3.5 billion years ago. Earth is the only place in the universe where life is known to exist. Some scientists believe that Earth-like biospheres might be rare.The biosphere is divided into a number of biomes, inhabited by broadly similar plants and animals. On land primarily latitude and height above the sea level separates biomes. Terrestrial biomes lying within the Arctic, Antarctic Circle or in high altitudes are relatively barren of plant and animal life, while the greatest latitudinal diversity of species is found at the Equator.

Natural Resources and Land Use

The Earth provides resources that are exploitable by humans for useful purposes. Some of these are non-renewable resources, such as mineral fuels, that are difficult to replenish on a short time scale.Large deposits of fossil fuels are obtained from the Earth's crust, consisting of coal, petroleum, natural gas and methane clathrate. These deposits are used by humans both for energy production and as feedstock for chemical production. Mineral ore bodies have also been formed in Earth's crust through a process of Ore genesis, resulting from actions of erosion and plate tectonics. These bodies form concentrated sources for many metals and other useful elements.

The Earth's biosphere produces many useful biological products for humans, including (but far from limited to) food, wood, pharmaceuticals, oxygen, and the recycling of many organic wastes. The land-based ecosystem depends upon topsoil and fresh water, and the oceanic ecosystem depends upon dissolved nutrients washed down from the land. Humans also live on the land by using building materials to construct shelters.

Natural and Environmental Hazards

Large areas are subject to extreme weather such as tropical cyclones, hurricanes, or typhoons that dominate life in those areas. Many places are subject to earthquakes, landslides, tsunamis, volcanic eruptions, tornadoes, sinkholes, blizzards, floods, droughts, and other calamities and disasters. Many localized areas are subject to human-made pollution of the air and water, acid rain and toxic substances, loss of vegetation (overgrazing, deforestation, desertification), loss of wildlife, species extinction, soil degradation, soil depletion, erosion, and introduction of invasive species.

A scientific consensus exists linking human activities to global warming due to industrial carbon dioxide emissions. This is predicted to produce changes such as the melting of glaciers and ice sheets, more extreme temperature ranges, significant changes in weather conditions and a global rise in average sea levels.

Human Geography

Earth has approximately 6,707,000,000 human inhabitants as of July 2008. Projections indicate that the world's human population will reach seven billion in 2013 and 9.2 billion in

2050. Most of the growth is expected to take place in developing nations. Human population density varies widely around the world, but a majority live in Asia. By 2020, 60% of the world's population is expected to be living in urban, rather than rural, areas. It is estimated that only one eighth of the surface of the Earth is suitable for humans to live on—three-quarters is covered by oceans, and half of the land area is either desert (14%), high mountains (27%), or other less suitable terrain. The northernmost permanent settlement in the world is Alert, on Ellesmere Island in Nunavut, Canada. (82°282 N) The southernmost is the Amundsen-Scott South Pole Station, in Antarctica, almost exactly at the South Pole. (90°S)

Independent sovereign nations claim all of the planet's land surface, with the exception of some parts of Antarctica. As of 2007 there are 201 sovereign states, including the 192 United Nations member states. In addition, there are 59 dependent territories, and a number of autonomous areas, territories under dispute and other entities. Historically, Earth has never had a sovereign government with authority over the entire globe, although a number of nation-states have striven for world domination and failed.

The United Nations is a worldwide intergovernmental organization that was created with the goal of intervening in the disputes between nations, thereby avoiding armed conflict. It is not, however, a world government. While the U.N. provides a mechanism for international law and, when the consensus of the membership permits, armed intervention, it serves primarily as a forum for international diplomacy. In total, about 400 people have been outside the Earth's atmosphere as of 2004, and, of these, twelve have walked on the Moon. Normally the only humans in space are those on the International Space Station. The station's crew of three people is usually replaced every six months.

CULTURAL VIEWPOINT

Etymology

The name *Earth* originates from the 8th century Anglo-

Saxon word *erda,* which means ground or soil. In Old English the word became *eorthe,* then *erthe* in Middle English. Earth was first used as the name of the sphere of the Earth around 1400. It is the only planet whose name in English is not derived from Greco-Roman mythology. The standard astronomical symbol of the Earth consists of a cross circumscribed by a circle. This symbol is known as the wheel cross, sun cross, Odin's cross or Woden's cross. Although it has been used in various cultures for different purposes, it came to represent the compass points, earth and the land. Another version of the symbol is a cross on top of a circle; a stylized globus cruciger that was also used as an early astronomical symbol for the planet Earth.

Religious Beliefs

Earth has often been personified as a deity, in particular a goddess. In many cultures the mother goddess, also called the Mother Earth, is also portrayed as a fertility deity. To the Aztec, Earth was called Tonantzin—"our mother"; to the Incas, Earth was called Pachamama—"mother earth". The Chinese Earth goddess Hou-T'u is similar to Gaia, the Greek goddess personifying the Earth. To Hindus it is called Bhuma Devi, the Goddess of Earth. In Norse mythology, the Earth goddess Jord was the mother of Thor and the daughter of Annar. Ancient Egyptian mythology is different from that of other cultures because Earth is male, Geb, and sky is female, Nut.

Creation myths in many religions recall a story involving the creation of the Earth by a supernatural deity or deities. A variety of religious groups, often associated with fundamentalist branches of Protestantism or Islam, assert that their interpretations of the accounts of creation in sacred texts are literal truth and should be considered alongside or replace conventional scientific accounts of the formation of the Earth and the origin and development of life. Such assertions are opposed by the scientific community and other religious groups. A prominent example is the creation-evolution controversy.

Exploration and Mapping

In the ancient past there were varying levels of belief in a flat Earth, with the Mesopotamian culture portraying the world as a flat disk afloat in an ocean. The spherical form of the Earth was suggested by early Greek philosophers; a belief espoused by Pythagoras. By the Middle Ages—as evidenced by thinkers such as Thomas Aquinas—European belief in a spherical Earth was widespread. Prior to circumnavigation of the planet and the introduction of space flight, belief in a spherical Earth was based on observations of the secondary effects of the Earth's shape and parallels drawn with the shape of other planets.

Cartography, the study and practice of map making, and vicariously geography, have historically been the disciplines devoted to depicting the Earth. Surveying, the determination of locations and distances, to a lesser extent navigation, the determination of position and direction, have developed alongside cartography and geography, providing and suitably quantifying the requisite information.

Modern Perspective

The technological developments of the latter half of the 20th century are widely considered to have altered the public's perception of the Earth. Before space flight, the popular image of Earth was of a green world. Science fiction artist Frank R. Paul provided perhaps the first image of a cloudless *blue* planet (with sharply defined land masses) on the back cover of the July 1940 issue of *Amazing Stories*, a common depiction for several decades thereafter. Earth was first photographed from space by Explorer 6 in 1959. Yuri Gagarin became the first human to view Earth from space in 1961.

The crew of the Apollo 8 was the first to view an Earth-rise from lunar orbit in 1968. In 1972 the crew of the Apollo 17 produced the famous "Blue Marble" photograph of the planet Earth from cislunar space. This became an iconic image of the planet as a marble of cloud-swirled blue ocean broken by green-brown continents. NASA archivist Mike Gentry has speculated that "The Blue Marble" is the most widely distributed image in human history. A photo taken of a distant Earth by *Voyager 1* in 1990 inspired Carl Sagan to describe the planet as a "Pale Blue Dot." Since the 1960s, Earth has also been described as a massive "Spaceship Earth," with a life support system that requires maintenance, or, in the Gaia hypothesis, as having a biosphere that forms one large organism.Over the past two centuries a growing environmental movement has emerged that is concerned about humankind's effects on the Earth.

The key issues of this socio-political movement are the conservation of natural resources, elimination of pollution, and the usage of land. Environmentalists advocate sustainable management of resources and stewardship of the environment through changes in public policy and individual behaviour. Of particular concern is the large-scale exploitation of non-renewable resources. Changes sought by the environmental movements are sometimes in conflict with commercial interests due to the additional costs associated with managing the environmental impact of those interests.

Future

The future of the planet is closely tied to that of the Sun. As a result of the steady accumulation of helium ash at the Sun's core, the star's total luminosity will slowly increase. The luminosity of the Sun will increase by 10 percent over the next 1.1 Gyr (1.1 billion years), and by 40% over the next 3.5 Gyr. Climate models indicate that the rise in radiation reaching the Earth is likely to have dire consequences, including the possible loss of the planet's oceans.

The Earth's increasing surface temperature will accelerate the inorganic CO_2 cycle, reducing its concentration to the lethal levels for plants (10 ppm for C4 photosynthesis) in 900 million years. The lack of vegetation will result in the loss of oxygen in the atmosphere, so animal life will become extinct within several million more years.

But even if the Sun were eternal and stable, the continued internal cooling of the Earth would have resulted in a loss of much of its atmosphere and oceans (due to lower volcanism). After another billion years the surface water will have completely disappeared and the mean global temperature will reach 70°C. The Earth is expected to be effectively habitable for another 500 million years or so.

The Sun, as part of its evolution, will expand to a red giant in about 5 Gyr. Models predict that the Sun will expand out to about 250 times its present size, roughly 1 AU (150,000,000 km). Earth's fate is less clear. As a red giant, the Sun will lose roughly 30% of its mass, so, without tidal effects, the Earth will be in an orbit 1.7 AU (250,000,000 km) from the Sun when the star reaches it maximum radius. Therefore, the planet is expected to escape envelopment by the expanded Sun's sparse outer atmosphere, though most, if not all, existing life will be destroyed because of the Sun's increased luminosity. However, a more recent simulation indicates that Earth's orbit will decay due to tidal effects and drag, causing it to enter the red giant Sun's atmosphere and be destroyed.

The Earth's shape is very close to an oblate spheroid—a rounded shape with a bulge around the equator—although the precise shape (the geoid) varies from this by up to 100 meters. The average diameter of the reference spheroid is about 12,742 km. More approximately the distance is 40,000 km/ð because the meter was originally defined as 1/10,000,000 of the distance from the equator to the north pole through Paris, France.

The rotation of the Earth creates the equatorial bulge so that the equatorial diameter is 43 km larger than the pole to pole diameter. The largest local deviations in the rocky surface of the Earth are Mount Everest (8,848 m above local sea level) and the Mariana Trench (10,911 m below local sea level). Hence compared to a perfect ellipsoid, the Earth has a tolerance of about one part in about 584, or 0.17%, which is less than the 0.22% tolerance allowed in billiard balls. Because of the bulge, the feature farthest from the centre of the Earth is actually Mount Chimborazo in Ecuador.

GEOLOGICAL OBSERVATION OF EARTH

The geologist is, above all things, an observer in the great outof-door world. The man whose horizon is bounded by the walls of a city can never be a geologist, though he may gain much scientific knowledge from books and from an inspection of collections in museums and laboratories. The true geologist, however, goes directly to the earth and there begins his inquiries. Not until observations of natural facts and phenomena were made in extenso was the inquiry of the philosophers regarding the earth and its history placed on a scientific basis.

Scattered observations and more or less accurate deductions were made even in antiquity. Thus Aristotle, in the third century B.C., had a very considerable understanding of the work of rivers and reasoned correctly regarding the changes in the land and sea at former times. The painter Leonardo da Vinci correctly reasoned, from an observation of the fossils found in the foothills of the Apennine Mountains, that they were the shells of once living animals, though they were generally regarded either as freaks of nature or as modern shells dropped by the pilgrims in their voyages across these mountains. It is true that the significance of fossil sea shells was recognized by the Greek philosophers but their explanations were generally ridiculed during the Middle Ages.

It was, however, not until the latter part of the eighteenth and the early part of the nineteenth century that systematic investigations of the rocks of the earth and their contained fossils began, and from this period we date the birth of geology as a science. Scientific geology arose more or less simultaneously in the different countries of Europe. In France Etienne Guettard and Nicholas Desmarest were among the first to bring observation of facts to the fore, while Buffon indulged in brilliant speculations on the origin of the earth. Later Alexander Brongniart investigàted the rocks around Paris, and Georges Cuvier described their fossils.

In Germany Abraham Gottlob Werner, who is often called the founder of German geology and who was Proessor at the Mining School at Freiberg, exerted a profound influence on geology especially by his teachings, to which men flocked from all countries. Being mostly an observer of the details of specimens and rarely venturing beyond his immediate surroundings for field observations, many of his geological deductions have proved erroneous—though his pupils and followers, notably Leopold von Buch, extended their observations over wide areas and added much to the store of geological facts as well as to its philosophy. Switzerland had its enthusiasticstudent of the structure and history of the Alps in the person of H. B. de Saussure, and Russia had in Pierre Simon Pallas its careful student of the Ural Mountains, and the rocks outcropping there and elsewhere in the empire. In Great Britain many men contributed to the discovery of facts and the interpretation of these.

Among them the first rank is given to James Hutton, whose work marks a turning point in the history of geology, for he insisted that "the past history of the globe must be explained by what can be seen happening now, or to have happened only recently," a dictum which has since become the very cornerstone of geology. Hutton great work, *Theory of the Earth with Proofs and Illustrations,* is better known through the classic volume, *Illustrations of the Huttonian Theory,* by his friend John Playfair, which no student of geology should neglect to read. In this work are contained many of the fundamental principles with which geologists are concerned to-day, and they are illustrated by a wealth of facts gleaned by Hutton from his rambles through Scotland and other countries. Another of the Scottish founders of geology was Sir James Hall, to whom we owe the origination of experimental geology.

The best known, however, among the early English geologists was William Smith, who is generally called the "Father of English Geology." He determined not only the correct successions of the English rock formations, and made the first geological map of England, but gave to many of the formations the names which they bear to-day. Finally, the student should remember among English geologists the name of Sir Charles Lyell, as that of a man who has had the most profound influence on geological thought. His great work, the *Principles of Geology,* has become a classic of geological literature. Among the men who exerted a profound influence on American geology in the early days of its development, the names of William McClure and Amos Eaton stand out prominently.

McClure, born in Scotland in 1763, became an American citizen near the close of the century. In 1809 he published the first important work on American Geology, in which appeared the first geological map of the Eastern United States, and one of the first geological maps of the country. Amos Eaton, born in New York state, is known especially for his *Index to the Geology of the Northern States,* which was the first geological textbook published in America, and in this and subsequent works he laid the foundation for the New York geological system. Many of the names of American formations still current were first applied by him. He also made the first geological map of New York. The important influence which McClure and Eaton had on the development of American geology has been recognized by the designation of the first two eras in the history of this science in America as the Maclurean and the Eatonian.

STUDY OF ROCKS AND ROCK STRUCTURES

The dwellers in the interior of our country, or the traveler on the broad plains of northern Germany, of Russia, of Hungary, or of China, will find little opportunity to get a view of the rocks which underlie these regions, for an almost continuous mantle of soil and drift covers the solid rock. Only where rivers have cut channels through the surface layer of loose material, the *mantlerock,* or where quarries have been opened, or construction operations have necessitated excavation down to and into the solid bed rock, is there an opportunity for observation of the rocks beneath the mantle-rock.

River valleys and gorges are therefore the favorite resorts of the geologist of the plains, while quarries, railroad cuts, and other excavations which expose the rock, likewise receive his attention. Borings and drillings for oil, gas, or water often prove of use to him, though generally, except where the diamond drill is used, and the core preserved, the record of such borings is only of doubtful and minor value. Salt shafts, such as those in central New York and that near Detroit, Mich., furnish excellent sections at the time of excavation.

Fortunately for the geologist of the plains, however, few regions of any great extent are wholly covered by mantle-rock; more commonly lowlying ridges of rock, formed by resistant layers, are exposed above the general surface of the soil and drift mantle. Such "ledges" generally mark the edges of beds of hard rock which have a gentle slope away from the edge of the ledge. When the ledge-forming layer is very thick, a cliff of some height may result, and this generally furnishes abundant opportunity for observation and deduction. All natural exposures of the rock are called *outcrops,* and the outcropping ledges together with the exposures in the stream channels, especially those which cut across the cliffs, furnish to the geologist of the flat countries his most satisfactory data. We may note a few examples. The state of New York furnishes an illustration of the types of rock outcrop in ledges referred to in the preceding paragraphs.

Over the greater part of the state the rocks are so gently inclined that they appear horizontal to the eye, and it is only when they are seen in the cliffs of Niagara gorge and along Lake Erie that their gentle southward descent becomes noticeable. Such a section, considerably generalized, is shown in the following diagram. Where the beds end in the air upon the north, the harder ones, such as limestones and sandstones, form a series of low cliffs, while the ends of the softer shale beds are usually marked by broad flat-bottomed valleys. The largest and deepest of these valleys is occupied by Lake Ontario, while others are filled by soil which conceals the rock.

These cliffs or escarpments generally have an abrupt northern face across the edge of the hard rock, and in these faces quarries are generally opened. These cliffs can be traced with more or less interruption across New York state from the Niagara River and Lake Erie to the Hudson, although they become modified because some of the beds seen in the western part of the state die out, or change in character and new ones appear. This is shown on the geological map of the state of New York, where a series of broad color bands extends east and west across the state. Each color band in general represents one rock layer or group of layers, and the width of the color band indicates the amount of exposure of each which would appear were all the covering soil and vegetation removed; or in other words, the amount by which each lower bed projects beyond the next higher one which covers it.

Wherever streams have cut across this series of beds, gorges are formed, in the walls of which the cut edges of the rocks, the soft layers as well as the hard ones, are shown. The most striking examples of such gorges are shown along the Genesee River, which crosses the state from south to north. In the section from Rochester northward it cuts the lower beds, the harder strata producing waterfalls. In the section between Portage and Mt. Morris, it cuts the higher strata, and here too several waterfalls are formed by hard layers. Between these two points many smaller tributary streams have cut into the sides of the valley and exposed the rocks.

Here, too, are situated several deep shafts which go down vertically to the Salina salt beds and during the cutting of which the succession of the rocky beds was ascertained. A good understanding of the succession of this series of rocky formations is obtained by the traveler who passes from the Adirondack Mountains southwestward across the state to Elmira, especially if he take advantage of the various sections exposed in the gorges of the streams and the banks of the Finger Lakes.

It was by the study of these natural outcrops of the state, supplemented by those made on the sections exposed during the cutting of the Erie Canal in 1817-1825, that the foundation of American geology was laid by such men as Amos Eaton, and by James Hall and others associated with him on the geological survey of New York state. The type of outcrop just described is found in most regions of flat-lying rocks in our own country as well as abroad. Thus the geologist who starts from Bamboo, Wisconsin, and proceeds southeastward to Lake Michigan at Milwaukee, will cross such a series of rocky ridges separated by soilfilled valleys, and a similar experience awaits the traveler across Kansas from southeast to northwest, or the one who proceeds southward across Oklahoma, or journeys from central Texas either northwestward or southeastward. The traveler across England from Liverpool to London also crosses such a series of rocky ridges, which in general extend in a northeast-southwest direction, and are formed by a succession of nearly horizontal rock-layers with westward facing cliffs, though these are not generally visible from the train.

Similar outcrops of nearly horizontal layers of rock surround Paris in constantly widening circles on three sides. The edges of some of these rocks form cliffs or escarpments facing outward. If one were to represent the rock formations which underlie Paris, and which have a shallow basin-like structure, by a nest of plates, the smallest at the top, the successive rims of these plates would represent the encircling cliffs, while the location of Paris would be in the center of the uppermost and smallest plate.

Many rivers have cut channels through the edges of these rock plates, while others flow in the depressions between successive rims. Thus numerous and excellent exposures of the rocks are found, even though the marginal portions between the edges of successive rock layers are frequently covered by soil and vegetation. These numerous rock exposures made possible the observations which gave the French founders of our science the data on which to build their deductions, and the cliffs which they form around Paris were of the utmost importance in the conduct of the Great War so recently closed. It is, however, in the elevated portions of the earth—the mountains and chains of uplands—that rock outcrops are most frequent on the surface of the land.

Here the soil and rock débris lodges only in the depressions, while between these the ledges protrude and give opportunities for observation. Here, too, deep cañons are cut by the rivers and glaciers and thus additional rock-walls are opened for observation. German, Austrian, and Swiss geologists have for the most part been limited to such regions for their observations, and the wonderful rock exposures of the Alps and other mountains have made it possible for them to carry their studies along certain lines to great lengths. Werner, the father of German geology, made most of his observations in the subdued mountain district of Saxony, especially the Erzgebirge, which has long been famous for its old and extensive mining operations. Since France and Italy also border on the Alps, and have mountain ranges of their own, the geologists of these countries were enabled to avail themselves of the rocks and rock structures thus revealed.

British geologists, too, have been able to some extent to resort to this type of exposure, though in these moisture-enveloped islands, as over parts of Scandinavia as well, the dense though low cover of vegetation and the peat accumulations obstruct much of the underlying rock, as all students of Irish geology know only too well. The Highlands of Scotland, however, furnish many good opportunities for observation, as do also many of the higher English districts and especially the mountain region of Wales.

Swedish geologists have frequently had to resort to digging through the surface layers to get at the underlying rock, and it is not an uncommon sight to see the Swedish geologist in the southern interior accompanied by a factotum, whose duty it is to wield pick and shovel. In European Russia and adjoining districts there are vast areas of flat-lying rocks covered by soil and drift, so that, except along the coast and in the river channels, outcrops of the bedrock are difficult to find. But where these rocks are uplifted—in the Ural Mountains on the east, the Caucasus on the south, and the Carpathians on the west—outcrops abound, and here the true relationships of the rock formations may be ascertained. In America, the New England uplands, the White, Green, and Adirondack Mountains, and the Appalachian Chain furnish an abundant series of rock outcrops for the eastern geologist.

The old and generally much-subdued mountain system, the rocks of which may be traced by frequent outcrops from New York City to the Highlands of the Hudson, and which can be followed southwestward through New Jersey, Pennsylvania, and Maryland, and indeed all the way down to the Carolinas, where it constitutes the older. Appalachian

Chain, is a typical example of an elevated, though for the most part not very mountainous, country, and here outcrops abound. Many of our western mountains are especially well adapted for geological observation, for here the aridity of the climate prevents the growth of much vegetation, and the rock structures, frequently developed on a gigantic scale, are visible for many miles.

This makes the Rocky Mountains a veritable paradise for the American geologist. Of all the natural rock exposures, however, those of the coast-line are the most attractive, and in many respects the most satisfying. Wherever the sea-coast or the shore of a large lake is formed by rocks which rise above the surface of the water, the cutting work of the waves keeps the exposure fresh. A tramp along a rocky sea-coast is replete with interest to the geologist, and many of the choicest bits of geological observation have been made on such sea-cliffs. Great Britain, with its wonderful rocky sea-coast, probably leads the world in the variety and significance of rocks and rock structures there exposed.

No student of geology can afford to neglect the wonderful English and Scottish coast, which has furnished the British geologists so many opportunities for the observation of facts that there, more than elsewhere, geological science has advanced, since the days of William Smith, with phenomenal strides. This is probably the reason why English geology quickly became the standard of comparison for other nations, in whose home countries observation was a more arduous task, because they did not include such marvelous coast exposures.

Northern France, too, has a coast-line of great interest to the geologist, and so has Norway. The coast-line of Germany, on the other hand, is mostly sandy, and there is little diversity in the types of the facts which it discloses. The Atlantic coast-line of North America is for the most part a sandy one. Only in New England, in the Canadian coastal prov inces, and in Newfoundland can be seen coastal sections comparable to some extent to those of Great Britain. The northern regions, however, are accessible with difficulty, and have only recently been investigated. But the New England coast, and especially that of Massachusetts, is a Mecca for American geologists, and many of the workers in American geology have had their preliminary training through a study of that interesting region.

Chapter 2

Volcano

Volcano, an opening in the earth's crust, through which heated matter is brought, permanently or temporarily, from the interior of the earth to the surface, where it usually forms a hill, more or less conical in shape, and generally with a hollow or crater at the top. This hill, though not an essential part of the volcanic mechanism, is what is commonly called the volcano. The name seems to have been applied originally to Etna and some of the Lipari Islands, which were regarded as the seats of Hephaestus, a Greek divinity identified with Vulcan, the god of fire in Roman mythology.

All the phenomena connected directly or indirectly with volcanic activity are comprised under the general designation of *vulcanism* or *vulcanicity - words* which are also written less familiarly as volcanism and volcanicity; whilst the study of the phenomena forms a department of natural knowledge known as *vulcanology*. Vulcanicity is the chief superficial expression of the earth's internal igneous activity.

It may happen that a volcano will remain for a long period in a state of moderate though variable activity, as illustrated by the normal condition of Stromboli, one of the Lipari Islands; but in most volcanoes the activity is more decidedly intermittent, paroxysms of greater or less violence occurring after intervals of comparative, or even complete, repose. If the period of quiescence has been very protracted, the renewed activity is apt to be exceptionally violent.

Thus, Krakatoa before the great eruption of 1883 had been dormant for something like two centuries, and it is believed that the Japanese volcano Bandaisan previously to the gigantic outburst of 1888 had been silent for more than a thousand years. A volcano may indeed remain so long dormant as to be mistaken for one completely extinct.

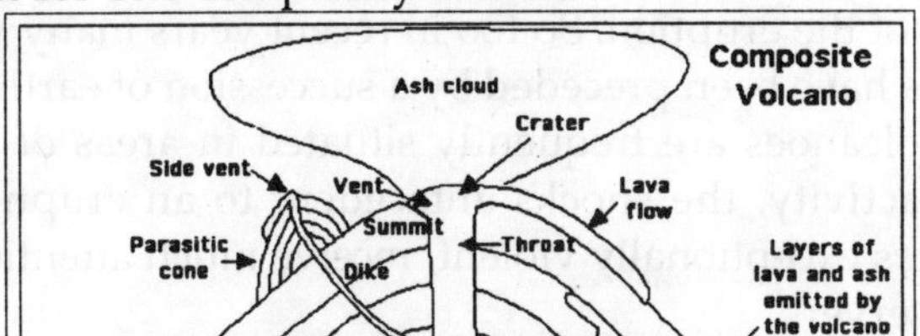

Fig. Features of Volcano

The volcanoes of central France are regarded as extinct, inasmuch as no authentic historical record of any eruption is known, but there are not wanting signs that in some parts of this volcanic region the subterranean forces may yet be slumbering rather than dead.

PREMONITORY SYMPTOMS

A volcanic eruption is usually preceded by certain symptoms, of which the most common are local earthquakes. The mountain, or other eruptive centre, may be thrown by internal activity into a state of tremor; the tremors perhaps continuing intermittently for months or even years, and becoming more frequent and violent as the crisis approaches. At first they are usually confined to the volcano and its immediate neighbourhood, but may subsequently extend to a considerable distance, though probably never developing into earthquakes of the first magnitude. The sudden opening of a subterranean crack, by rupture of a rock under strain, or the rapid injection of lava into such a fissure, will tend to produce a jar at the surface. For at least sixteen years before the first recorded eruption of Vesuvius in A.D. 79 earthquakes had been frequent in the Campania and had wrought havoc in the cities of Herculaneum and Pompeii. Again, the formation of Monte Nuovo, near Pozzuoli, in 1538, was heralded by local earthquakes beginning several years in advance of the eruption. So too in recent years many volcanic outbursts have been preceded by a succession of earthquakes; but as volcanoes are frequently situated in areas of marked seismic activity, the shocks antecedent to an eruption may not, unless exceptionally violent, receive much attention from local observers.

It commonly happens that a volcanic outburst is announced by subterranean roaring and rumbling, often compared to thunder or the discharge of artillery underground. Other precursory symptoms may be afforded by neighbouring springs, which not unusually flow with diminished volume, or even fail altogether. Possibly fissures open underground and drain off the water from the springs and wells in the immediate locality. Occasionally, however, an increased flow has been recorded. In some cases thermal springs make their appearance, whilst the temperature of any existing warm springs may be increased, and perhaps carbon dioxide be evolved. A disturbed state of the atmosphere is by no means a constant forerunner of an eruption, some of the greatest outbursts having occurred in a period of atmospheric stability: indeed the air is often felt to be close and still.

Immediately before a renewed outburst in an old volcano, the floor of the crater is generally upheaved to a greater or less extent, whilst the discharge of vapour from any fumaroles is increased. Where a crater has been occupied by water, forming a crater-lake, the water on the approach of an eruption becomes warm, evolves visible vapour, and may even boil. In the case of cones which are capped with snow, the internal heat of the rising lava usually causes a rapid melting of the snow-cap, resulting perhaps in a disastrous deluge.

It seems probable that by attention to the premonitory symptoms a careful local observer might in many cases foretell an eruption.

It generally happens that a great eruption is preceded by a preliminary phase of feeble activity. Thus, the gigantic catastrophe at Krakatoa on the 27th of August 1883, so far from having been a sudden outburst, was the culmination of a state of excitement, sometimes moderate and sometimes violent, which had been in progress for several months.

Emission of Vapour

Of all volcanic phenomena the most constant is the emission of vapour. It is one of the earliest features of an eruption; it persists during the paroxysms, attaining often to prodigious volume; and it lingers as the last relic of an outburst, so that long after the ejection of ashes and lava has ceased an occasional puff of vapour may be the only memento of the disturbance.

By far the greatest proportion of the vapour is steam, which sometimes occurs almost to the exclusion of other gaseous products. Such, at least, is the usual and probably correct view, though it is opposed by A. Brun, who regards the volcanic vapours as chiefly composed of chlorides with steam in only subordinate amount. In the case of a mild eruption, like that occurring normally at Stromboli, the vapours may be discharged in periodical puffs, marking the explosion of bubbles rising more or less rhythmically from the seething lava in the volcanic cauldron. S. Wise observed at the volcano of Sangay, in Ecuador, no fewer than 267 explosions in the course of an hour, the vapour here being associated, as is so often the case, with ashes.

During a violent eruption the vapour may be suddenly shot upwards as a vertical column of enormous height, penetrating the passing clouds. For a short distance above the vent the superheated steam sometimes exists as a transparent vapour, but it soon suffers partial condensation, forming clouds, which, if not dispersed by winds, accumulate over the mountain. When the vapour is free from ash it forms rolling balls of fleecy cloud, but usually it carries in mechanical association more or less finely divided lava as volcanic dust and ashes, whereby it becomes yellow, brown, or even black, sometimes as foul as the densest smoke. In a calm atmosphere the dust-laden vapour may rise in immense rings with a rotatory movement, like that of vortex-rings. Frequently the vapours, emitted in a rapid succession of jets, form cumulus clouds, or are massed together in cauliflower-like forms. The well-known "pine-tree appendage" of Vesuvius, noted by the younger Pliny in his first letter to Tacitus on the eruption in the year 79, is a vertical shaft of vapour terminating upwards in a canopy of cloud, and compared popularly with the trunk and spreading branches of the stonepine.

Whilst in some cases the cloud resembles a gigantic expanded umbrella, in others it is more mushroom-shaped. In a great eruption, the height of the mountain itself may appear dwarfed by comparison with that of the column of vapour.

During the eruption of Vesuvius in April 1906, the steam and dust rose to a height of between 6 and 8 m. At Krakatoa in 1883 the column of vapour and ashes reached an altitude of nearly 20 m.; whilst it was estimated by some authorities that during the most violent explosions the finely divided matter must have been carried to an elevation of more than 30 m. The emission of vast volumes of vapour at high tension naturally produces much atmospheric disturbance, often felt at great distances from the centre of eruption.

ELECTRICAL EXCITEMENT

It is probably to the uprushing current of vapour that much of the electrical excitement which invariably accompanies an eruption may be referred. The friction of the steam rushing in jets through the volcanic vent must produce electrical disturbance, and indeed an active volcano has been aptly compared to a hydroelectric machine of gigantic power. Another cause of excitement may be found in the mutual friction of the ejected cinders and ashes as they rise and fall in showers through the air. Much trituration of volcanic material may go on in the crater and elsewhere during the eruption, whereby the solid lava is reduced to a fine dust.

Other means of generating electricity are found in the chemical reactions effected in the volcano and in the sudden condensation of the emitted vapour. L. Palmieri, in the course of his investigations at the observatory on Vesuvius, found that the vapours free from cinders carried a positive charge, whilst the cinders were negative.

The electrical phenomena attending an eruption are often of great intensity and splendour. The dark ash-laden clouds of vapour are shot through and through by volcanic lightning, sometimes in rapid horizontal flashes, then in oblique forked streaks, or again in tortuous lines compared to fiery serpents, whilst the borders of the cloud may be brilliant with electric scintillations, often forming balls and stars of fire. During the great eruption of Krakatoa remarkable phenomena were observed by ships in the Strait of Sunda, luminous balls like "St Elmo's fire" appearing at the mast-heads and the yard-arms, whilst the volcanic mud which fell upon rigging and deck was strongly phosphorescent.

Quite distinct from any electrical phenomena is that intermittent reddish glare which is often seen at night in clouds hanging over an active crater, and which is simply a glow due to reflection from,the incandescent lava and stones in the volcanic cauldron below.

VOLCANIC RAIN AND MUD

The condensation of the vast volumes of steam exhaled during an eruption produces torrents of rain, which, mingling to a greater or less extent with the volcanic ashes, forms a hot muddy stream known in Italy as *lava d'acqua* and *lava di fango,* and in South America as *moya.* Deluges of such mud-lava may rush violently down the mountain-side and spread over the neighbouring country with terribly destructive effect, whence they are greatly dreaded by those who dwell at the base of a volcano.

The solidified volcanic mud, often mingled with larger fragments of lava, is known as *duff* or *tufa.* Herculaneum was buried beneath a flood of mud swept down from Vesuvius during the Plinian eruption of 79, and the hard tufaceous crust which thus sealed up the illfated city came in turn to be covered by lava-flows from subsequent eruptions: hence the difficulty of excavating at Herculaneum compared with similar work at Pompeii, where there was probably much less mud, since the city, having been at a greater distance from the volcanic centre, was overwhelmed in great measure by loose ashes, capable of removal with comparative ease.

Fig. Volcanic Mud

It sometimes happens that volcanic mud is formed by the mingling of hot ashes not directly with rain but with water from streams and lakes, or even, as in Iceland, with melted snow. A torrent of mud was one of the earliest symptoms of the violent eruption of Mont Pele in Martinique in 1902. This mud had its source in the Etang Sec, a crater-basin high up on the S.W. side of the mountain.

By the explosive discharge of ashes and vapours mingled with the water of the tarn there was produced a vast volume of hot muddy matter which on the 5th of May suddenly escaped from the basin, when a huge torrent of boiling black mud, charged with blocks of rock and moving with enormous rapidity, rolled like an avalanche down the gorge of the Riviere Blanche. If a stream of lava obstructs the drainage of a volcano, it may give rise to floods.

EJECTED BLOCKS

When a volcano after a long period of repose starts into fresh activity, the materials which have accumulated in the crater, including probably large blocks from the disintegration of the crater-walls, have to be ejected. If the lava from the last eruption has consolidated as a plug in the throat of the volcano, the conduit may be practically closed, and hence the first effort of the renewed activity is to expel this obstruction. The hard mass becomes shattered by the explosions, and the angular fragments so formed are hurled forth by the outrushing stream of vapour. When the discharge is violent, the vapour, as it rushes impetuously up the volcanic duct, may tear fragments of rock from its walls and project them to a considerable distance from the vent. Such ejected blocks, by no means uncommon in the early stages of an eruption, are often of large size and naturally vary according to the character of the rocks through which the duct has been opened.

They may be irregular masses of igneous rocks, possibly lavas of earlier eruptions, or they may be stratified, sedimentary and fossiliferous rocks representing the platform on which the volcano has been built, or the yet more deeply seated fundamental rocks. By Dr H. J. Johnston-Lavis, who specially studied the ejected blocks of Vesuvius, the volcanic materials broken from the cone are termed "accessory" ejecta, whilst other fragmentary materials he conveniently calls "accidental" products, leaving the term "essential" ejecta for plastic lava, ashes, crystals etc.

Masses of Cretaceous or Apennine limestone ejected from Somma are scattered through the tuffs on the slopes of Vesuvius; and objects carved in such altered limestone are sold to tourists as "lava" ornaments. Under the influence of volcanic heat and vapours, the ejected blocks suffer more or less alteration, and may contain in their cavities many crystallized minerals. Certain blocks of sandstone ejected occasionally at Etna are composed of white granular quartz, permeated with vitreous matter and encased in a black scoriaceous crust of basic lava.

A rock consisting of an irregular aggregation of coarse ejected materials, including many large blocks, is known as a "volcanic agglomerate." Any fragmental matter discharged from a volcano may form rocks which are described as "pyroclastic." *Cinders, Ashes and* Dust. - After the throat of a volcano has been cleared out and a free exit established, the copious discharge of vapour is generally accompanied by the ejection of fresh lava in a fragmentary condition. If the ejected masses bear obvious resemblance to the products of the hearth and the furnace, they are known as "cinders" or "scoriae," whilst the small cinders not larger than walnuts often pass under their Italian name of "lapilli". When of globular or ellipsoidal form, the ejected masses are known as " bombs " or "volcanic tears".

Other names are given to the smaller fragments. If the lava has become granulated it is termed "volcanic sand"; when in a finer state of division it is called ash, or if yet more highly comminuted it is classed as dust; but the latter terms are sometimes used interchangeably. The pulverized material, consisting of lava which has been broken up by the explosion, or triturated in the crater, is often discharged in prodigious quantity, so that after an eruption the country for miles around the volcano may be covered with a coating of fine ash or dust, sometimes nearly white, like a fall of snow, but often of greyish colour, looking rather like Portland cement, and in many cases becomirg reddish by oxidation of the ferruginous constituents.

Even when first ejected the ash is sometimes cocoa-coloured. This finely divided lava insinuates itself into every crack and cranny, reaching the interior of houses even when windows and doors are closed. A heavy fall of ash or cinders may cause great structural damage, crushing the roofs of buildings by sheer weight, as was markedly the case at Ottajano and San Guiseppe during the eruption of Vesuvius in April 1906. On this occasion the dry ashes slipped down the sides of the volcanic cone like an avalanche, forming great ashslides with ridges and furrows rather like barrancos, or ravines, caused by rain. The burial of Ottajano and San Ginseppe in 1906 by Vesuvian ejecta, mostly lapilli, has been compared with that of Pompeii in 79.

Deposits of volcanic sand and ashes retain their heat long after ejection, so that rain will cause them to evolve steam, and if the rain be heavy and sudden it may produce explosions with emission of great clouds of vapour. The fall of ash is at first prejudicial to vegetation, and is often accompanied or followed by acid rain; but ultimately the ash may prove beneficial to the soil, chiefly in consequence of the alkalis which it contains. The "May dust" of Barbados was a rain of volcanic ash which fell in May 1812 from the eruption of the Soufriere in St Vincent. It is estimated that the amount of dust which during this eruption fell on the surface of Barbados, Ioo m. distant from the eruptive centre, was about 3,000,000 tons. The distance to which ash is carried depends greatly on the atmospheric conditions at the time of the eruption. Ashes from Vesuvius in an eruption in the year 472 were carried, it is said, as far as Constantinople.

During an eruption of Cotopaxi, on the 3rd of July 1880, observed by E. Whymper, an enormous black column of dust-laden vapour was shot vertically upwards with such rapidity that in less than a minute it rose to a height estimated at 20,000 ft. above the crater-rim, or nearly 40,000 ft. above sea-level, when it was dispersed by the wind over a very wide area. It is believed that the amount of dust in this discharge must have been more than 2,000,000 tons.

Enormous quantities of dust ejected from Krakatoa in 1883 were carried to prodigious distances, samples having been collected at more than a thousand miles from the volcano; whilst the very fine material in ultramicroscopic grains which remained suspended for months in the higher regions of the atmosphere seems to have enjoyed an almost world-wide distribution, and to have been responsible for the remarkable sunsets at that period.

The ash falling in the immediate vicinity of a volcanic vent will generally be coarser than that carried to a distance, since the particles as they are wafted through the air undergo a kind of sifting. Professor J. W. Judd, who made an exhaustive examination of the products of the eruption of Krakatoa, found that the dust near the volcano was comparatively coarse, dense and rather darkcoloured, in consequence of the presence of numerous fragments of heavy, dark, crystalline minerals, whilst the dust at a distance was excessively fine and perfectly white. According to this observer, the particles tended to fall in the following order: magnetite, pyroxenes, felspar, glass. The finely comminuted material, carried to a great height in the atmosphere, consisted largely of delicate threads and attenuated plates of vitreous matter, in many cases hollow and containing air-bubbles. The greater part of the dust was formed by the mutual attrition of fragments of brittle pumice as they rose and fell in the crater, which thus became a powerful "dust-making mill." By this trituration of the pumiceous lava, carried on for a space of three months during which the eruption lasted, the quantity of finely pulverized material must have been enormous; yet the amount of ejected matter was probably very much less than that extruded during some other historical eruptions, such as that of Tomboro in Sumbawa, in 1815.

The explosions at Krakatoa were, however, exceptionally violent, having been sufficient to project some of the finely pulverized lava to an altitude estimated to have been at least 30 m. It is usually impossible during a great eruption to determine the height of the column of "smoke" since it hangs over the country as a pall of darkness.

The great black cloud, which was so characteristic a feature in the terrible eruptions in the West Indies in 1902, was formed of steam with sulphur dioxide and other gases, very heavily charged with incandescent sand or dust, forming a dense mixture that in some respects behaved like a liquid. Unlike the Krakatoa dust, which was derived from a vitreous pumice, the solid matter of the black cloud was largely composed of fragments of crystalline minerals. According to Drs Anderson and Flett it is not impossible that on the afternoon of the 17th of May 1902, the solid matter ejected from the Soufriere of St Vincent amounted to several billions of tons, and that some of the dust fell at distances more than 2000 m. east of the centre of eruption.

LAVA

The volcanic cinders, sand, ashes and dust described above are but varied forms of solidified lava. Lava is indeed the most characteristic product of volcanic activity. It consists of mineral matter which is, or has been, in a molten state; but the liquidity is not due to simple dry fusion. The magma, or subterranean molten matter, may be regarded as composed essentially of various silicates, or their constituents, in a state of mutual solution, and heavily charged with certain vapours or gases, principally water-vapour, superheated and under pressure. In consequence of the peculiar constitution of the magma, the order in which minerals separate and solidify from it on cooling does not necessarily correspond with the inverse order of their relative fusibility.

The lava differs from the magma before eruption, inasmuch as water and various volatile substances may be expelled on extrusion. The rapid escape of vapour from the lava contributes to the explosive phenomena of an eruption, whilst the rate at which the vapour is disengaged depends largely on the viscosity of the magma.

The lava on its immediate issue from the volcanic vent is probably at a white heat, but the temperature is difficult of determination since the molten matter is usually not easy of approach, by reason of the enshrouding vapour. Determinations of temperature are generally made at a short distance from the exit, when the lava has undergone more or less cooling, or on a small stream from a subordinate vent. A.

Bartoli, using a platinum electric resistance pyrometer, found that a stream of lava near a *bocca,* or orifice of emission, on Etna, in the eruption of 1892, had at a depth of one foot a temperature of 1060° C. In the lavas of Vesuvius and Etna thin wires of silver and of copper have frequently been melted. Probably the lava at the surface of the stream has a temperature of something like 110o° C., but this must not be assumed to be its temperature at the volcanic focus. C. Doelter, in some experiments on the melting-point of lava by means of an electric furnace, found that a lava from Etna softened at from 962° to 970° C. and became fluid at 1010° to 1040°, whilst a Vesuvian lava softened at 1030° to 1060° and acquired fluidity at 1080° to 1090°.

These results were obtained at ordinary atmospheric pressure, but it has been assumed that the melting-point of lava at a great depth would, through pressure alone, exceed that obtained in the laboratory. On the other hand the presence of water and of certain volatile fluxes in the magma lowers the fusing-point, and hence the extruded lava from which these have largely escaped may be much less fusible than the original magma.

Fig. Volcanic Lava

It should be noted that all determinations of the melting-points of minerals and rocks involving ocular inspection of the physical state of the material are liable to considerable error, and the only accurate method seems to be that of determining the point at which absorption of heat abruptly occurs - the latent heat of fusion. It is believed that the temperature of lava in the volcanic conduit may be in some cases sufficiently high to fuse the neighbouring rocks, and so melt out a passage through them in its ascent. The wallrock thus dissolved in the magma will not be without influence on the composition of the lava with which it becomes assimilated.

Many interesting observations are on record with regard to the heating effect of lava on metals and other objects with which it may have come in contact. Thus, after the destruction of Torre del Greco by a current of lava from Vesuvius in 1794, it was found that brass in the houses under the lava had suffered decomposition, tje copper having become crystallized; whilst silver had been not only fused but sublimed. This indicates a temperature of upwards of 1000° C. Panes of glass in the windows at Torre del Greco on the same occasion suffered devitrification.

Notwithstanding the high temperature of lava on

emission, it cools so rapidly, and the consolidated lava conducts heat so slowly, that vegetable structures may be involved in a lava-flow without being entirely destroyed. A stream of lava on entering a wood, as in the sylvan region on Etna, may burn up the undergrowth but leave many of the larger trees with their trunks merely carbonized. On Vesuvius a lava-flow has been observed to surround trees while the foliage has been apparently uninjured.

A vertical trunk of a coniferous tree partially enveloped in Tertiary basalt occurs at Gribon in the Isle of Mull, as described by Sir A. Geikie and others; plant-remains in basalt from the Bo'ness coalfield in Linlithgowshire have been noticed by H. M. Cadell; and attention has been called by B. Hobson to a specimen of scoriaceous basalt, from Mexico, which shows the impression of ears of maize and even relics of the actual grains. In consequence of the slow transmission of heat by solid lava, the crust on the surface of a stream may be crossed with impunity whilst the matter is still glowing at a short distance below. Lichens may indeed grow on lava which remains highly heated in the interior. The solidified surface of a sheet of lava may be smooth and shining, sometimes quite satiny in sheen, though locally wrinkled and perhaps even ropy or hummocky, the irregularities being mainly due to superficial movement after partial solidification.

The "corded lava" has a surface similar to that often seen on blastfurnace slag, and is suggestive of a tranquil flow. After a lava stream has become crusted over on cooling, the subjacent lava, still moving in a viscous condition, tends to tear the crust, forming irregular blocks, or clinkers, which are carried forward by the flow and ultimately left in. the form of confused heaps, perhaps of considerable magnitude. The front of a stream may present a wall of scoriaceous fragments looking like a huge pile of coke. As the clinkers are carried along, on the surface of the lava, they produce by mutual friction a crunching noise; and the sluggish flow of the lava-stream laden with its burden has been compared with that of a glacier.

Since the upper part of the stream moves more rapidly than the lower, which is retarded by cooling in contact with the bed-rock, the superficial clinkers are carried forward and, rolling over the end, may become embedded in the lava as it advances. Scoriae formed on the top of a stream may thus find their way to the base. Rockfragments or other detrital matter occurring in the path of the lava will be caught up by the flow and become involved in the lower part of the molten mass; whilst the rocks over which the lava travels may suffer more or less alteration by the heat of the stream.

The rapidity of a lava flow is determined partly by the slope of the bed over which it moves and partly by the consistency of the lava, this being dependent on its chemical composition and on the conciitions of cooling. In an eruption of Mauna Loa, in Hawaii, in 1855, the lava was estimated to flow at a rate of 40 m. an hour; and at an eruption of Vesuvius in 1805 a velocity of more than 50 m. an hour, at the moment of emission, was recorded.

The rapidity of flow is, however, rapidly checked as the stream advances, the retardation being very marked in small flows. Where lava travels down a steep incline there is naturally a great tendency to form a rugged surface, whilst a quiet flow over a flat plane favours smoothness. If the lava meet a precipice it may form a cascade of great beauty, the clinkers rapidly rolling down with a clatter, as described by Sir W. Hamilton in the eruption of Vesuvius in 1771, when the fiery torrent had a perpendicular fall of 50 ft.

In Hawaii the smooth shining lava, often superficially waved and lobed, is known as *pahoehoe,* whilst the rugged clinker beds are termed aa. These terms are now used in general terminology, having been introduced by American geologists. The fields of aa often contain lava-balls and bombs. It may be said that the pahoehoe corresponds practically with the *Fladen lava* of German vulcanologists, and the as with their *Schollen lava.* Rugged flows are known in Auvergne as *cheires.* The surface of a clinker-field has often a horribly jagged character, being covered with ragged blocks bristling with sharp points. In the case of an obsidian-flow a most dangerous surface is produced by the keen edges and points of the fragmentary volcanic glass.

If, after a stream of lava has become crusted over, the underlying magma should flow away, a long cavern or tunnel may be formed. Should the flow be rapid the roof may collapse and the fragments, falling on to the stream, may be carried forward or become absorbed in the fused mass. The walls and roof of a lava-cave are occasionally adorned with stalactites, whilst the floor may be covered with stalagmitic deposits of lava. The volcanic stalactites are slender, tubular bodies, extremely fragile, often knotted and rippled.

Beautiful examples of lava stalactites from Hawaii have been described by Professor E. S. Dana. Caverns may also be formed in lava-flows by the presence of large bubbles, or by the union of several bubbles. It may happen, too, that certain monticules thrown up on the surface of the lava are hollow, of which a famous example is furnished by the Caverne de Rosemond, at the base of Piton Barry, in the Isle of Reunion.

Fig. Lava Arch, Hawaii

It is of great interest to determine whether molten lava contracts or expands on solidification, but the experimental evidence on this subject is rather conflicting. According to some observers a piece of solid lava thrown on to the surface of the same lava in a liquid state will sink, while according to others it floats. It has often been observed that cakes formed by the natural fracture of the crust on the lava of Kilauea sink in the liquid mass, but it has been suggested that the fragments are drawn down by convection-currents.

On the other hand a solid piece, though denser than the corresponding liquid, may be buoyed up for a time by the viscous condition of the molten lava. Moreover, the presence of minute vesicles may lighten the mass. Although the minerals of a rock-magma may separately contract on crystallization it does not follow that the magma itself, in which they probably exist in a state of solution, will undergo on crystallization a similar change of volume. On the whole, however, there seems reason to believe that lava on solidifying almost always diminishes in volume and consequently increases in density.

According to the experiments of C. Doelter the specific gravity of molten lava is invariably less than that of the same lava when solid, though in some cases the difference is but slight. In a vitreous or isotropic condition the lava has a lower density than when crystalline.

The differences are illustrated by the following table, where the figures give the specific gravity: Experiments by Dr C. Barus showed that a diabase of specific gravity 3.017 formed a glass of sp. gr. 2.717, and melted to a liquid of sp. gr. 2.52. J. A. Douglas on examining various igneous rocks found that in all cases the rock in a vitreous state had a lower sp. gr. than in a crystalline condition, the difference being greatest in the acid plutonic rocks. A. Harker, however, has called attention to the fact that the glassy selvage of certain basic dykes in Scotland is denser than the same rock in a crystalline condition in the interior of the dykes.

PHYSICAL STRUCTURE OF LAVAS

An amorphous vitreous mass may result from the rapid cooling of a lava on its extrusion from the volcanic vent. The common type of volcanic glass is known as obsidian. Microscopic examination usually shows that even in this glass some of the molecules of the magma have assumed definite orientation, forming the incipient crystalline bodies known as microlites. By the increase of these minute enclosures, in number and magnitude, the lava may become devitrified and assume a lithoidal or stony structure. If the molten magma consolidate slowly, the various silicates in solution tend to separate by crystallization as their respective points of saturation are reached. Should the process be arrested before the entire mass has crystallized, the crystals that have been developed will be embedded in the residual magma, which may, on consolidation, form a vitreous base. It is believed that in many cases the lava brings up, through its conduit, myriads of crystals that have been developed during slow solidification in the heart of the volcanic apparatus. Showers of crystals of leucite have occurred at Vesuvius, of labradorite at Etna, and of pyroxene at Vesuvius, Etna and Stromboli. These intratelluric crystals were probably floating in the molten magma, and had they remained in suspension, this magma might on consolidation have enveloped them as a ground-mass or base. A rock so formed is generally known as a "porphyry" and the structure as porphyritic.

In such a lava the large crystals, or phenocrysts, evidently represent an early phase of consolidation, and the minerals of the matrix a later stage. It is notable that the intratelluric crystals often lack sharpness of outline, as though they had suffered corrosion by attack of the molten magma, whilst they may contain vitreous enclosures, suggesting that the surrounding mass was liquid during their consolidation. It is believed that the more slowly consolidation has occurred, the larger generally are the crystals; and the higher the temperature of the magma the greater the corrosion or resorption. Possibly under certain conditions the phenocrysts and the ground-mass may have solidified simultaneously.

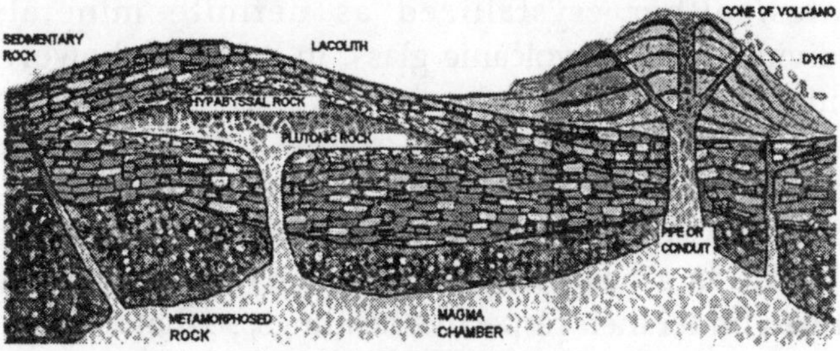

Fig. Structure of Lavas

In some cases the entire igneous mass assumes a crystalline structure, or becomes "holocrystalline". Such a structure is well displayed when the magma has consolidated at considerable depths, cooling slowly under great pressure, and forming rocks which are termed "plutonic" or "abyssal" to distinguish them from rocks truly volcanic, or those which, if not effusive, like lava-flows, have at least solidified very near to the surface as dykes and sills. Volcanic and plutonic rocks pass, however, into each other by gradual transition. The dyke-rocks, or intrusive masses, form an intermediate group sometimes distinguished under the name of "hypabyssal" rocks, as suggested by W. C. Brogger. Lavas extruded in submarine eruptions may have solidified under a great weight of seawater, and therefore to that extent rather under plutonic conditions.

CHEMICAL COMPOSITION OF LAVAS

Lavas are usually classified roughly, from a chemical point of view, in broad groups according to the proportion of silica which they contain. Those in which the proportion of silica reaches 66% or upwards are said to be acid or acidic, whilst those in which it falls to 55% or below are called basic lavas. The two series are connected by a group of intermediate composition, whilst a small number of igneous rocks of exceptional type are recognized as ultrabasic. Professor F. W. Clarke has suggested a grouping of igneous rocks as per-silicic, medio-silicic and subsilicic, in which the proportion of silica is respectively more than 60, between 50 and 60, or less than 50%.

By far the greater part of all lavas consists of various silicates, either crystallized as definite minerals or unindividualized as volcanic glass. In addition, however, to the mineral silicates, a volcanic rock may contain a limited amount of free acid and basic oxides, represented by such minerals as quartz and magnetite. Rhyolite may be cited as a typical example of an acid lava, andesite as an intermediate and basalt as a basic lava. The various volcanic rocks are described under their respective headings, so that it is needless to refer here to their chemical or mineralogical composition.

In the course of the life of a volcano, the lava which it emits may undergo changes, within moderate limits, being at one time more acid, at another more basic. Such changes are sometimes connected with a shifting of the axis of eruption. Thus at Etna the lavas from the old axis of Trifoglietto in the Valle del Bove were andesites, with about 55% of silica, but those rising in the present conduit are doleritic, with a silica-content of only about 50%. It seems probable that, to a limited extent, changes in the character of a lava may sometimes be due to contact of the magma with different rocks underground: if these are rich in silica, the acidity of the lava will naturally increase; while if they are rich in calcareous and ferromagnesian constituents, the basicity will increase: the variation is consequently apt to be only local, and probably always slight.

By von Richthofen and some others it has been held that during a long period of igneous activity a definite order in the succession of the erupted rocks is everywhere constant; but though some striking coincidences may be cited, it can hardly be said that this generalization has been satisfactorily established. It has, however, often been observed, as emphasized by Professor Iddings, that a volcanic centre will start with the emission of lavas of neutral or intermediate type, followed in the course of a geological period by acid and basic lavas, and ending with those of extreme composition, indicating progressive change in the magma.

The old idea of a universal magma, or continuous pyrosphere, has been generally abandoned. Whatever may have been the case in a primitive condition of the interior of the earth, it seems necessary to admit that the magma must now exist in separate reservoirs. The independent activity of neighbouring volcanoes strikingly illustrated in Kilauea and Mauna Loa in Hawaii, only 20 m. apart, suggests a want of communication between the conduits; and though the lavas are very similar at these two centres, it would seem that they can hardly be drawn from a common source. Again, the volcanoes of southern Italy and the neighbouring islands exhibit little or no sympathy in their action, and emit lavas of diverse type. The lavas of Vulcano, one of the Lipari Isles, are rhyolitic, whilst those of Stromboli, another of the group, are basaltic.

It is believed that the magma in a subterranean reservoir, though originally homogeneous, may slowly undergo certain changes, whereby the more basic constituents migrate to one quarter whilst the acid segregate in another, so that the canal, at successive periods, may bring up material of different types. It has often been observed that all the rocks from a definite igneous centre have a general similarity in chemical and mineralogical characters. This relationship is called, after Professor Iddings, "consanguinity" and appears to be due to the fact that the rocks are drawn from a common source. Professor Judd pointed out the existence of distinct "petrographical provinces", within which the eruptive rocks during a given geological period have a certain family likeness and have appeared in definite succession. Thus he recognized a Brito-Icelandic petrographical province of Tertiary and recent lavas. It has been shown by A. Harker that alkali igneous rocks are generally associated with the Atlantic type of coast-line and sub-alkali rocks with the Pacific type.

Although changes in the character of an erupted product from a given centre are usually brought about very slowly, it has often been supposed that even in the course of a single prolonged eruption, or series of eruptions, the character of the lava may vary to some extent. That this is not, however, usually the case has been repeatedly proved. M. H. Arsandaux, for instance, analysed the bombs of augite-andesite thrown out from Santorin at the beginning of the eruption of 1866, others ejected in 1867, and others again at the close of the eruption in 1868; and he found no important variation in the composition of the magma during these successive stages. Moreover, Professor A. Lacroix found that the material extruded from Vesuvius in 1906 remained practically of the same composition from the beginning to the end of the eruption, and further, that it presented great analogy to that of 1872 and even to that of 1631.

All the Vesuvian lavas are of the type of rock known as leucotephrite or leucitetephrite, or they pass, by the presence of a little olivine, into leucite-basanite. Leucite is characteristic of the lavas of Vesuvius, whilst it is excessively rare in those of Etna, where a normal doleritic type prevails. Nepheline, a felspathoid related to leucite, is characteristic of certain lavas, such as those of the Canary Islands, which comprise nephelinetephrites and nepheline-basanites. Most of the lavas from the volcanoes of South America consist of hypersthene-andesite, and it is notable that the fragmental ejectamenta from the eruptions of St Vincent and Martinique in 1902 and from Krakatoa in 1883 were evidently derived from a magma of this Pacific type.

It commonly happens that acid lavas are paler in colour, less dense and less fusible than basic lavas, and they are probably drawn in some cases from shallower depths. As a consequence of the ready fusibility of many basic lavas, they flow freely on emission, running to great distances and forming far-spreading sheets, whilst the more acid lavas rapidly become viscid and tend to consolidate nearer to their origin, often in hummocky masses. The shape of a volcanic mountain is consequently determined to a large extent by the chemical character of the lavas which it emits. In the Hawaiian Islands, for instance, where the lavas are highly basic and fluent, they form mountains which, though lofty, are flat domes with very gently sloping sides.

Such is the fluidity of the lava on emission that it flows freely on a slope of less than one degree. In consequence, too, of this mobility, it is readily thrown into spray and even projected by the expansive force of vapour into jets, which may rise to the height of hundreds of feet and fall back still incandescent, producing the appearance of "fire fountains." The emission is not usually accompanied, however, by violent explosions, such as are often associated with the eruption of magmas of less basic and more viscous nature. The viscosity of the lava at Kilauea was estimated by G. F. Becker to be about fifty times as great as that of water. It may be pointed out that the fusibility of a lava depends not on the mere fact that it is basic, but rather on the character of the bases. A lava from Etna or Vesuvius may be really as basic as one from Hawaii.

Capillary Lava

A filamentous form of lava well known at Kilauea, in Hawaii, is termed *Pele's hair,* after Pele, the reputed goddess of the Hawaiian volcanoes. It resembles the capillary slag much used in the arts under the name of " mineral wool " - a material formed by injecting steam into molten slag from an iron blast-furnace. It is commonly supposed that Pele's hair has been formed from drops of lava splashed into the air and drawn out by the wind into fine threads.

According, however, to Major C. E. Dutton, the filaments are formed on the eddying surface of the lava by the elongation of minute vesicles of water-vapour expelled from the magma. C. F. W. Krukenberg, who examined the hair microscopically, figured a large number of fibres, some of which showed the presence of minute vesicles and microscopic crystals, the former when drawn out rendering the thread tubular. In a spongy vitreous scoria from Hawaii, described as " thread-lace," a polygonal network of delicate fibres forms little skeleton cells. Capillary lava is not confined to the Hawaiian volcanoes: it is known, for example, in Reunion, and may be formed even at Vesuvius.

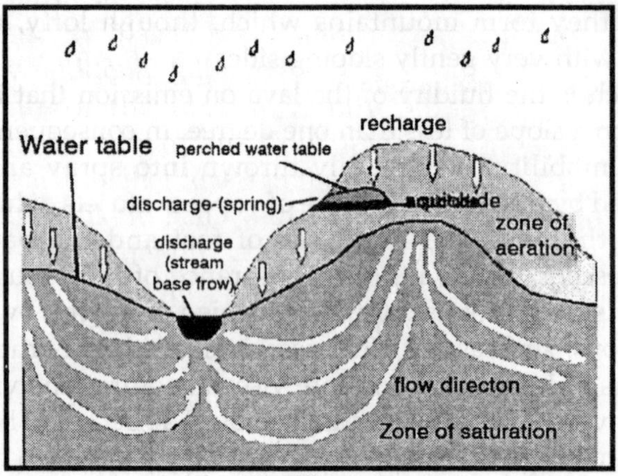

Fig. Capillary Lava

Pumiceous Lava

The copious disengagement of vapour in a glassy lava gives rise to the light cellular or spongy substance, full of microscopic pores, known as pumice. It is usually, though not invariably, produced from an acid lava, and may sometimes be regarded as the solidified foam of an obsidian. During the eruption of Krakatoa in 1883 enormous quantities of pumice were ejected, and were carried by the sea to vast distances, until they ultimately became water-logged and sank. Professor Judd found the pumice to consist of a vitreous lava greatly inflated by imprisoned vapours; the walls of the air-cells were formed of the lava drawn out into thin plates and threads, often with delicate fibres running across the cavities.

Having been suddenly cooled, it was extremely brittle, and its ready pulverization gave rise to much of the ash ejected during this eruption. It has been shown by Dr JohnstonLavis that a bed of pumiceous lava, especially if basic, is generally vitreous towards the base, becoming denser, darker and more crystalline upwards, until it may pass superficially into scoria. The change is explicable by reduction in the temperature of the magma consequent on the conversion of water into steam.

Water in Lavas

Whether an eruption is of an explosive or a tranquil character must depend largely, though not wholly, on the chemical composition of the magma, especially on the extent to which it is aquiferous. By relief of pressure on the rise of the column in the volcanic channel, or otherwise, more or less steam will be disengaged, and if in large quantity this must become, with other vapours, a projectile agency of enormous power. The precise physical condition in which water exists in the magma is a matter of speculation, and hence Johnston-Lavis proposed to designate it simply as H 2 O. Water above its critical point, which is about 370° C. or 698°F., cannot exist as a liquid, whatever be the pressure, neither is it an ordinary vapour. It has been estimated that the critical point would probably be reached at a depth of about 7 m. At very high temperatures the elements of water may exist in a state of dissociation.

Much discussion has arisen as to the origin of the volcanic water, but probably it is not all attributable to a single source. Some may be of superficial origin, derived from rain, river or sea; whilst the upward passage of lava through moist strata must generate large volumes of steam. It has often been remarked that wet weather increases the activity of a volcano, and that in certain mountains the eruptions are more frequent in winter. According, however, to Professor A. Ricco's prolonged study of Etna, rain has no apparent influence on the activity of this mountain, and indeed the number of eruptions in winter, when rains are abundant, seems rather less than in summer.

Fig. Lava in The ocean

The popular belief that explosive action is due to the admission of water to the volcanic focus is founded mainly on the topographic relation of volcanoes to large natural bodies of water, many being situated near the shore of a continent or on islands or even on the sea-floor. Salt water gaining access to heated rocks, through fissures or by capillary absorption, would give rise not only to watervapour but to the volatile chlorides so common in volcanic exhalations.

Yet it is notable that comparatively little chlorine is found among the products exhaled by the volcanoes of Hawaii, though these are typically insular. L. Palmieri, however, described certain sublimates on lava at Vesuvius after the eruption of 1872 as deposits of "sea-salt," to show that they were not simply sodium chloride, but contained other constituents found in sea-water.

It has been supposed that water on reaching the hot walls of a subterranean cavity would pass into the spheroidal state, and on subsequent reduction of temperature might come into direct contact with the heated surface, when it would flash with explosive violence into steam. Such catastrophes probably occur in certain cases. When, for example, a volcano becomes dormant, water commonly accumulates in the crater, and on a renewal of activity this craterlake may be absorbed through fissures in the floor leading to the reopened duct, and thus become rapidly, even suddenly, converted into vapour. But such incidents are accidental rather than normal, and seem incompetent to account for volcanic activity in general.

When a stream of lava flows into the sea it no doubt immediately generates a prodigious volume of steam; but this is only a temporary phenomenon, for the lava rapidly becomes chilled by the cold water, with formation of a superficial solid layer, which by its low thermal conductivity allows the internal mass to cool slowly and quietly.

In the great eruption of Krakatoa in 1883 the sea-water gained occasional access to the molten lava, and by its cooling effect checked the escape of vapour, thus temporarily diminishing the volcanic activity. But Judd compares this action to that of fastening down the safety-valve of a steam-boiler. The tension of the elastic fluids being increased by this repression would give rise subsequently to an explosion of greater violence; and hence the short violent paroxysms characteristic of the Krakatoa eruption were due to what he calls a "check and rally" of the subterranean forces. The action in the volcanic conduit has, indeed, been compared with that of a geyser.

The downward passage of water through fissures must be confined to the upper portion of the earth's crust known as the " zone of fracture," for it is there only that open channels can exist. Water might also percolate through the pores of the rocks, but even the pores are closed at great depths. It was shown many years ago by G. A. Daubree that water could pass to a limited extent through a heated rock against the pressure of steam in the opposite direction. According to S. Arrhenius, water may pass inwards through the sea-bottom by osmotic pressure.

As the melting points of various silicates are lowered by admixture with water, it appears that the access of surface-waters to heated rocks must promote their fusibility. Judd has suggested that the proximity of large bodies of water may be favourable to volcanic manifestations, because the hydrated rocks become readily melted by internal heat and thus yield a supply of lava.

Whilst some of the water-vapour exhaled from a volcano is undoubtedly derived from superficial sources, notably in such insular volcanoes as Stromboli, the opinion has of late years been gaining ground, through the teaching of Professor E. Suess and others, that the volcanic water must be largely referred to a deep-seated subterranean origin - that it is, in a word, "hypogene" or magmatic rather than meteoric. It is held that the magma as it rises through the volcanic conduit brings up much water-vapour and other gaseous matters derived from original sources, perhaps a relic of what was present in the earth in its molten condition, having possibly been absorbed from a dense primordial atmosphere, or, as suggested by Professor T. C. Chamberlin, entrapped by the globe during its formation by accretion of planetesimal matter.

Water brought from magmatic depths to the surface, and appearing there for the first time, has been termed " juvenile," and it has been assumed that such water may be seen in hot springs like those at Carlsbad. Professor J. W. Gregory has suggested that certain springs in the interior of Australia may derive part of their supply from juvenile or plutonic waters. According to A. Gautier, the origin of volcanic water may be found in the oxidation of hydrogen, developed from masses of crystalline rock, which by subsidence have been subjected to the action of subterranean heat.

Volcanic Vapours

It seems not unlikely that the vapours and gases exist in the volcanic magma in much the same way that they can exist in molten metal. It is a familiar fact that certain metals when melted can absorb large volumes of gases without entering into chemical combination with them. Molten silver, for example, is capable of absorbing from the atmosphere more than twenty times its volume. of oxygen, which it expels on solidification, thus producing what is called the "spitting of silver." Platinum again can absorb and retain when solid, or occlude, a large volume of hydrogen, that can be expelled by heating the metal in vacuo. In like manner molten rock under pressure can absorb much steam. It appears that many igneous rocks contain gases locked up in their pores, not set free by pulverization, yet capable of expulsion by strong heat.

Fig. Volcanic Vapours

Tilden has found that granite, gabbro, basalt and certain other igneous rocks enclose many times their volume of gases, chiefly hydrogen and carbon dioxide, with carbon monoxide, methane and nitrogen. Thus, the basalt of Antrim in Ireland, which is a Tertiary lava, yielded eight times its volume of gas having the following percentage composition: hydrogen 36.15, carbon dioxide 32.08, carbon monoxide 20.08, methane 10, nitrogen. No doubt some of the gases evolved on heating rocks may be generated by reactions during the experiment, as shown by M. W. Travers, and also by Armand Gautier.

It has been pointed out by Gautier that the gas exhaled from Mont Pele during the eruption of 1902 had practically the same composition as that which he obtained on heating granite and certain other rocks. According to this authority a cubic kilometre of granite heated to redness would yield not less than 26,000,000 tons of water-vapour, besides other gases.

If then a mass of granite in the earth's crust were subject to a great local accession of heat it might evolve vast volumes of gaseous, matter, capable of producing an eruption of explosive type. Judd found that the little balls of Siberian obsidian called marekanite threw off, when strongly heated, clouds of finely divided particles formed by rupture of the distended mass through the escape of vapour. Pitchstone when ignited loses in some cases as much as to % of its weight, due to expulsion of water.

Much of the steam and other vapour brought up from below by the lava may be evolved on mere exposure to the air, and hence a stream freshly extruded is generally beclouded with more or less vapour. Gaseous bubbles in the body of the lava render it vesicular,. especially in the upper part of a stream, where the pressure is relieved, and the vesicles by the onward flow of the lava tend to become elongated in the direction of movement.

Vesiculation, being naturally resisted by cohesion, is not common in very viscid lavas of acid type, nor is it to be expected where the lava has been subject to great pressure, but it is seen to perfection in surface-flows of liquid lavas of basaltic character. A vesicular structure may sometimes be seen even in dykes, but the cavities are usually rounded rather than elongated, and are often arranged in bands parallel to the walls of the dyke. A very small proportion of water in a lava. suffices to produce vesiculation. Secondary minerals developed in a cellular lava may be deposited in the steam-holes, thus producing an amygdaloidal rock.

After the surface of a lava-stream has become crusted over, vapour may still be evolved in the interior of the mass, and in seeking release may elevate or even pierce the crust. Small cones may thus be thrown up on a lava-flow, and when vapour escapes from terminal or lateral orifices they are known as " spiracles." The steam may issue with sufficient projectile force to toss up the lava in little fountains. When the lava is very liquid, as in the Hawaiian volcanoes, it may after projection from the blow-hole fall back in drops and plastic clots, which on consolidation form, by their union, small cones.

Vapour-vents on lava are often known as fumaroles. The character of the gaseous exhalations varies with the temperature, and the following classification was suggested by C. Sainte-Claire Deville:

- Dry or white fumaroles having a temperature above

500° C. and evolving compounds of chlorine, and perhaps fluorine.

- Acid fumaroles, exhaling much steam, with hydrocholoric acid and sulphur dioxide.
- Alkaline fumaroles, at a temperature of about 100°, with much steam and ammonium chloride and some sulphuretted hydrogen.
- Cold fumaroles, below100°, with aqueous vapour, carbon dioxide and sulphuretted hydrogen.
- Mofettes, indicating the expiring phase of vulcanism.

A similar sequence of emanations, following progressive cooling of the lava, has been noted by other observers. During an eruption, the gaseous products may vary considerably. Johnston-Lavis found at Vesuvius that the vapour which first escaped from the boiling lava contained much sulphurous acid, and that hydrochloric acid and other chlorides. appeared later.

If the vapours exhaled from volcanoes were derived originally from superficial sources, the lava would, of course, merely return to the surface of the earth what it had directly or indirectly absorbed. But if, as is now rather generally believed, much if not most of the volcanic vapour is derived from original subterranean sources, it must form a direct contribution from the interior of the earth to the atmosphere and hydrosphere, and consequently becomes of extreme geological interest.

DESCRIPTION OF SPECIAL GASES AND VAPOURS

Hydrochloric acid, HCI, escapes abundantly from many vents, often accompanied with the vapours of certain metallic chlorides, and is responsible for much of the acrid effects of volcanic exhalations. To avoid dangerous vapours an active volcano should be ascended on the windward side. Free hydrofluoric acid, HF, has sometimes been detected with the hydrochloric acid among Vesuvian vapours, and silicon fluoride, SiF 4, has also been reported. Sulphuretted hydrogen, H 2 S, is a frequent emanation, and being combustible may contribute to the lambent flames seen in some eruptions. It readily suffers oxidation, giving rise to sulphur dioxide and water. By the interaction of hydrogen sulphide and carbon dioxide, water and carbon oxysulphide, COS, are formed; whilst by reaction with sulphur dioxide, water and free sulphur are produced, such being no doubt the origin of many deposits of volcanic sulphur.

Hydrogen sulphide may be formed by the decomposition of certain metallic sulphides, like that of calcium, in the presence of moisture, as suggested by Anderson and Flett with regard to certain muds at the Soufriere of St Vincent. Sulphur dioxide, SO_2, is one of the commonest exhalations, especially at acid fumaroles. It may be detected by its characteristic smell, that of burning brimstone, even when present in very small proportion and in the presence of an excess of hydrochloric acid. By hydration it readily forms sulphurous acid, which may be further oxidized to sulphuric acid. J. B. Boussingault found free sulphuric acid (with hydrochloric acid) in the water of the Rio Vinagre which issues from the volcano of Purace in the Andes of Colombia; and it occurs also in certain other volcanic waters.

Carbon dioxide, CO_2, is generally a product of the later stages of an eruption and is often evolved after all other gases have ceased to escape. Although it may sometimes be due to the decomposition of limestone, it seems to be mostly of true magmatic origin. At the well-known Grotta del Cane, at Lake Agnano, in the Phlegraean Fields near Naples, there has been for ages a copious discharge, and analyses of the air of the cave by T. Graham Young showed the presence of from 61.5 to 7 1% of carbon dioxide. Gautier, in 1907, found 96 to 97% of this gas in the vapours (excluding water-vapour) emitted from the Solfatara near Pozzuoli in the Bay of Naples. The gas by its density tends to accumulate in depressed areas, as in the Death Gulch in the Yellowstone Park and in the Upas Valley of Java. In the Eifel, in the Auvergne and in many other volcanic regions it is discharged at temperatures not above that of the atmosphere.

This natural carbonic acid gas is now utilized industrially

at many localities. In the gases of the fumaroles of Mont Pele, carbon monoxide, CO, was detected by H. Moissan. Probably certain hydrocarbons, notably methane or marsh-gas, CH4, often exist in volcanic gases. They might be formed by the action of water on natural carbides, such as that of magnesium, calcium. Moissan found 5.46% of methane in vapour from a fumarole on Mont Pele in 1902.

Free hydrogen was detected by R. Bunsen as far back as 1846 in vapours from volcanoes in Iceland. In 1861 Deville and Fouque found it, with hydrocarbons, at Torre del Greco near Naples; and in 1866 Fouque discovered it at Santorin, where some of the vapour at the immediate focus of eruption contained as much as 30% of hydrogen. It is notable that at Santorin free oxygen was also found. The elements of water may possibly exist, at the high temperature of the magma, in a state of dissociation, and certain volcanic explosions have sometimes been attributed to the combination of these elements.

Oxygen is not infrequently found among volcanic emanations, but may perhaps be derived in most cases from superficial air and ground-water; and in like manner the nitrogen, often detected, may be sometimes of atmospheric origin, though in other cases derived from nitrides in the lava. In the vapours emitted by Mont Pele in 1902 argon was detected by H. Moissan, to the extent of 0.71%; and in those from Vesuvius in 1906 argon and neon were found by Gautier. The collection of volcanic vapours offers difficulty, and it is not easy to avoid admixture with the atmosphere. F. A. Perret has successfully collected gases on,Vesuvius.

VOLCANIC FLAMES

Although the incandescence of the lava and stones projected during an eruption, and the reflection from incandescent matter in the crater have often been mistaken for red flames, there can be no doubt that true combustion, though generally feeble, does occur during volcanic outbursts. Among the gases cited above, hydrogen, hydrogen sulphide and the hydrocarbons are inflammable. The flames seen in volcanoes are generally pale and of bluish, greenish or yellowish tint. The presence of hydrogen, sodium and hydrocarbons in the volcanic flames of Kilauea. These, however, were referred to the kindling of sulphur deposited around the fumaroles, the flames being coloured by the presence of boric acid and arsenic sulphide.

When a stream of lava flows over vegetation the combustion of the leaves and wood may be mistaken for flames issuing from the lava. In like manner brushwood may grow in the crater of a dormant volcano and be ignited by a fresh outburst of lava, thus producing flames which, from their position in the crater, may readily deceive an observer.

Volcanic Sublimates

Certain mineral substances occur as sublimates in and around the volcanic vents, forming incrustations on the lava. They are either deposited directly from the effluent vapours, which carry them in a volatile condition, or are produced by interaction of the vapours among themselves; whilst some of the incrustations, rather loosely called sublimates, are due to reaction of the vapours on the constituents of the lava. Possibly at the temperature of the magma-reservoirs even silica and various silicates may be volatilized, and might thus yield sublimation products. Many of the volcanic sublimates occur at first as incandescent crusts on the lava. Being generally unstable they are difficult of preservation, and are not usually well represented in collections.

Among the commonest sublimates is halite, or sodium chloride, NaCl, occurring as a white crystalline incrustation, sometimes accompanied, as at Vesuvius, by sylvite, or potassium chloride, KCl, which forms a similar sublimate. The two chlorides may be intimately associated. Sal ammoniac, or ammonium chloride, NH_4 C1, is not uncommon, especially at Etna, as a white crystalline crust, probably formed in part by the reaction of hydrochloric acid with nitrogen and hydrogen in the vapours.

Bunsen, on finding it in Iceland, regarded it as a product of the distillation of organic matter. At the Solfatara, near

Pozzuoli, sal ammoniac was formerly collected as a sublimate on tiles placed round a bocca or vapourvent. Ferric chloride, FeCl 3, not infrequently occurs as a reddish or brownish yellow deliquescent incrustation, and because it thus colours the lava it has received the name of molysite.

The action of hydrochloric acid on the iron compounds in the lava may readily yield this chloride, which from its yellowish colour has sometimes been mistaken for sulphur. A crystalline sublimate from the fumaroles on Vesuvius, containing ferric and alkaline chlorides, KCl NH_4 Cl $2FeC1_3$ + $6H_2O$, is known as kremersite, after P. Kremers. From a scoriaceous lava found on Vesuvius after the eruption of 1906, Johnston-Lavis procured a yellow rhombohedral sublimate, which he proved to be a chloride of manganese and potassium, whence he proposed for it the name chlormanganokalite. It was studied by L. J.

Spencer, and found to contain 4KC1. MnC12. Chlorocalcite, or native calcium chloride, $CaC1_2$, has been found in cubic crystals on Vesuvian lava. Fluorite, or calcium fluoride, CaF_2, is also known as a volcanic product. Lead chloride, $PbCl_2$, a rare Vesuvian mineral, was named cotunnite, after Dr Cotugno of Naples. The action of hydrogen sulphide on this chloride may give rise to galena, PbS, found by A. Lacroix on Vesuvius in 1906. Atacamite, or cupric oxychloride, CuC1 2.3Cu (OH)2, occurs as a green incrustation on certain Vesuvian lavas, notably those of 1631. Another green mineral from Vesuvius was found by A. Scacchi to be a sulphate containing copper, with potassium and sodium, which he named from its fine colour *euclorina* - a word which has been written in English as euchlorinite.

The copper in the sublimates on Vesuvius will sometimes plate the iron nails of a traveller's boots when crossing the newly erupted lava. Cupric oxide, CuO, occurs in delicate crystalline scales termed tenorite, after Professor G. Tenore of Naples; whilst cupric sulphide, CuS, forms a delicately reticulated incrustation known as covellite, after N. Covelli, its discoverer at Vesuvius.

A sublimate not infrequently found in feathery crystalline

deposits on lava at Vesuvius, and formerly called " Vesuvian salt," is a potassium and sodium sulphate, (K. Na) 2 SO_4, known as aphthitalite (from Gr. a4B%Tos, imperishable, and &As, salt). A sulphate with the composition PbS04 (K Na)2 SO_4, found in the fumaroles at Vesuvius after the eruption of 1906, was named by A. Lacroix palmierite, after L. Palmieri, who was formerly director of the observatory on Vesuvius. Various sulphites are formed on lavas by the sulphurous acid of the vapours.

Ferric oxide, Fe_2O_3, which occurs in beautiful metallic scales as specular iron-ore, or as an amorphous reddish incrustation on the lava, is probably formed in most cases by the interaction of vapour of ferric chloride and steam at a high temperature. Less frequently, magnetite, Fe304, and magnesioferrite, MgFe204, are found in octahedral crystals on lava. An iron nitride was detected thinly incrusting a lava erupted at Etna in 1874, and was named by 0. Silvestri, who examined it, siderazote.

Boric acid, H3B03, occurs in the crater of Vulcano so abundantly that it was at one time collected commercially. It has also led to the foundation of an industry in Tuscany, where it is obtained from the soffioni of the Maremma. From Sasso in Tuscany it has received the name of sassolin or sassolite. Realgar, or arsenic sulphide, As 2 S 2, occurs in certain volcanic exhalations and is deposited as an orange-red incrustation, often associated with sulphur, as at the Solfatara, where orpiment, As_2S_3, has also been found.

Of all volcanic products, sulphur is in some respects the most important. It may, in large quantity lining the walls of the crater, as at Popocatepetl in Mexico, where it was formerly worked by the Indian " volcaneros," or on the other hand it may be a rare product, as at Vesuvius. Sulphur appears generally to owe its origin in volcanic areas to the interaction of sulphur dioxide and sulphuretted hydrogen, or to the action of water on the latter.

A volcanic vent where sulphur is deposited is truly a solfatara (*solfo terra*) or a soufriere, but all volcanoes which have passed into that stage in which they emit merely heated vapours now pass under this name. The famous Solfatara, an old crater in the Phlegraean Fields, exhales sulphurous vapours, especially at the Bocca Grande, from which sulphur is deposited. In the orangecoloured sulphur of the Solfatara, realgar may be present to the extent of as much as 18%. A brown seleniferous sulphur occurring at Vulcano, one of the Lipari Islands, was termed by W. Haidinger volcanite, but it should be noted that Professor W. H. Hobbs has applied this name to an anorthoclase-augite rock ejected as bombs at Vulcano. Sulphur containing selenium is known as a volcanic product in Hawaii, whilst in Japan not only selenium but tellurium occurs in certain kinds of sulphur.

At the Solfatara, near Pozzuoli, the hot sulphurous vapours attack the trachytic rocks from which they issue, giving rise to such products as alum, kaolin and gypsum. To some of these products, including alunogen and mendozite (soda-alum), the name solfatarite was given by C. W. Sheppard in 1835. By prolonged action of the acid vapours on lava, the bases of the silicates may be removed, leaving the silica as a soft white chalk-like substance. The occurrence of kaolin and other white earthy alteration-products led to the hills around the Solfatara being known to the Romans as the *Colli leucogei.*

THE HOT DUST CLOUD AND AVALANCHE OF PELE

The terrific eruptions in the islands of Martinique and St Vincent in the West Indies in 1902, furnished examples of a type of activity not previously recognized by vulcanologists, though, as Professor A. Lacroix has pointed out, similar phenomena have no doubt occurred elsewhere, especially in the Azores. By Drs Tempest Anderson and J. S. Flett, who were commissioned by the Royal Society to report on the phenomena, this type of explosive eruption is distinguished as the " Pelean type."

Its distinctive character is found in the sudden emission of a dense black cloud of superheated and suffocating gases, heavily charged with incandescent dust, moving with great velocity and accompanied by the discharge of immense volumes of volcanic sand, which are not rained down in the normal manner, but descend like a hot avalanche. The cloud, with the avalanche, is called by Lacroix a *nuee Peleenne,* or *nuee ardente,* the latter term having been applied to the fatal cloud in the eruptions at San Jorge in the Azores in 1818. In its typical form, the cloud seen at Pele appeared as a solid bank, opaque and impenetrable, but having the edge in places hanging like folds of a curtain, and apparently of brown or purplish colour. Rolling along like an inky torrent, it produced in its passage intense darkness, relieved by vivid lightning. So much solid matter was suspended in the cloud, that it became too dense to surmount obstacles and behaved rather like a liquid.

It has, however, been suggested that its peculiar movement as it swept down the mountain was due not simply to its heavy charge of solids, but partly to the oblique direction of the initial explosion. After leaving the crater, it underwent enormous expansion, and Anderson and Flett were led to suggest that possibly at the moment of emission it might have been partly in the form of liquid drops, which on solidifying evolved large volumes of gas held previously in occlusion. The deadly effect of the blast seems to have been mostly due to the irritation of the mucous membrane of the respiratory passages by the fine hot dust, but suffocating gases, like sulphur dioxide and sulphuretted hydrogen, were associated with the water-vapour. Possibly the incandescent dust was even hotter than the surrounding vapour, since the latter might be cooled by expansion.

It is said that the black cloud as it swept along was accompanied by an indraught of air, not however sufficiently powerful to check its rapid advance. The current of air was likened by Anderson and Flett to the inrush of air at a railway station as an express train passes. An attempt was made to determine the temperature of the fatal blast which destroyed St Pierre, but without very definite results. Thus it was assumed that as the telephone wires were not melted the temperature was below the fusing-point of copper: possibly, however, the blast may have passed too rapidly to produce the effects which might normally be due to its temperature.

SHAPE OF VOLCANIC CONES

Those volcanic products which are solid when ejected, or which solidify after extrusion, tend to form by their accumulation around the eruptive vent a hill, which, though generally more or less conical, is subject to much variation in shape. It occasionally happens that the hill is composed wholly of ejected blocks, not themselves of volcanic origin. In this case an explosion has rent the ground, and the effluent vapours have hurled forth fragments of the shattered rock through which the vent was opened, but no ash or other fragmentary volcanic material has been ejected, nor has any lava been poured forth.

This exceptional type is represented in the Eifel by certain monticules which consist mainly of fragments of Devonian slate, more or less altered. In some cases the area within a ring of such rocky materials is occupied by a sheet of water, forming a crater-lake, known in the Eifel as a *rnaar*. Piles of fragmentary matter of this character, though containing neither cinders nor lava, may be fairly regarded as volcanic, inasmuch as they are due to the explosive action of hot subterranean vapours.

In the ordinary paroxysmal type of eruption, however, cinders and ashes are shot upwards by the explosion and then descend in showers, forming around the orifice a mound, in shape rather like the diminutive cone of sand in the lower lobe of an hour-glass. Little cindercones of this character may be formed within the crater of a large volcano during a single eruption; whilst large cones are built up by many successive discharges, each sheet of fragmentary material mantling more or less regularly round the preceding layer.

The symmetry of the hill is not infrequently affected by disturbing influences - a strong wind, for example, blowing the loose matter towards one side. The sides of a cinder cone have generally a steep slope, varying from 300 to 45°, depending on the angle of repose of the ejectamenta. Excellent examples of small scoria-cones are found among the puys of Auvergne in central France, whilst a mag nificent illustration of this type of hill is furnished by Fusiyama, in Japan, which reaches an altitude of 12,000 ft. How such a cone may be rapidly built up was well shown by the formation of Monte Nuovo, near Pozzuoli - a hill 400 ft. high and a mile and a half in circumference, which is known from contemporary evidence to have been formed in the course of a few days in September 1538. The shape of a cinder cone may be retained for ages, since it is not liable to suffer greatly by denudation, as the rain soaks into the loose porous mass instead of running down the outside. If lava rises in the duct of a cinder cone, it may, on accumulation in the crater, break down the wall, and thus effect its escape as a stream. Cones breached in this way are not uncommon in Auvergne.

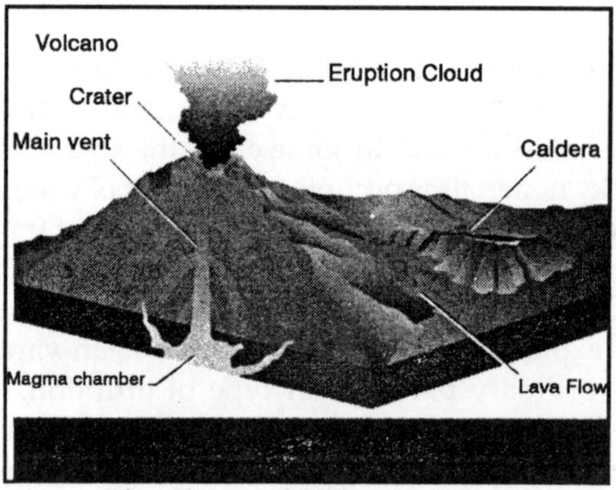

Fig. Shape of Volcanic Cones

It often happens that the cinders and ashes ejected from a volcano become mixed with water, and so form a paste, which sets readily as a hard tufaceous mass. Such natural tuff is indeed similar to the hydraulic cement known as pozzolana, which is formed artificially from volcanic ashes, and is renowned for durability. Although streams of volcanic mud are commonly associated with the ashes of a cinder-cone they may also form independent structures or tuffcones. These are generally broad-topped hills, having sides with an angle of slope as low in some cases as 15°.

Lava-cones are built up of streams of lava which have consolidated around the funnel of escape. Associated with the lava, however, there is usually more or less fragmentary matter, so that the cones are composite in structure and consequently more acute in shape than if they were composed wholly of lava. As the streams of lava in a volcano run at different times in different directions, they radiate from the centre, or flow from lateral or eccentric orifices, as irregular tongues, and do not generally form continuous sheets covering the mountain.

When lava is the sole or chief element in the cone, the shape of the hill is determined to a great extent by the chemical composition and viscosity of the lava, its copiousness and the rapidity of flow. If the lava be highly basic and very mobile, it may spread to a great distance before solidifying, and thus form a hill covering a large area and rising perhaps to a great height, but remarkably flat in profile. Were the lava perfectly liquid, it would indeed form a sheet without any perceptible slope of surface. As a matter of fact, some lavas are so fluent as to run down an incline of 1 °, and flat cones of basalt have in some cases a slope of only 10° or even less.

The colossal mass of Mauna Loa, in Hawaii, forms a remarkably flat broad cone, spreading over a base of enormous area and rising to a height of 13,900 ft. Major Dutton, writing in 1883, said that " a; moderate eruption of Mauna Loa represents more material than Vesuvius has emitted since the days of Pompeii." Yet the lava is so mobile that it generally wells forth quietly, without explosive demonstration, and therefore unaccompanied by fragmentary ejectamenta. Fluent lavas like those of Hawaii are also poured forth from the volcanoes and volcanic fissures of Iceland.

If the lava be less basic and less fusible, the hill formed by its accumulation instead of being a low dome will take the shape of a cone with sides of higher gradient: in the case of andesite cones, for instance, the slope may vary from 25° to 35°. Acid rocks, or those rich in silica, such as rhyolites and trachytes, may be emitted as very viscous lavas tending to form dome-shaped or bulbous masses.

Experiment shows that such lavas may persist for a considerable time in a semi-solid condition. It is possible for a dome to increase in size not by the lava running over the crater and down the sides but by injection of the pasty magma within the expanding bulb while still soft; or if solidified, the crust yields by cracking. Such a mode of growth, in which the dome consists of successive sheets that have been compared to the skins of an onion, has been illustrated by the experiments of Dr A. Reyer, and the structure is typically represented by the mamelons or steep-sided domes of the Isle of Bourbon.

The Puy-de-Dome in Auvergne is an example of a cone formed of the trachytic rock called from its locality domite, whilst the Grand Sarcoui in the same region illustrates the broad domeshaped type of hill. Such domes may have no summit-crater, and it is then usually assumed that the top with the crater has been removed by denudation, but possibly in some cases such a feature never existed. The " dome volcano " of von Seebach is a dome of acid lava extruded as a homogeneous mass, without conspicuous chimney or crater. Although domes are usually composed of acid rocks, it seems possible that they may be formed also of basic lavas, if the magma be protruded slowly at a low temperature so as to be rapidly congealed.

The Spine of Pele

A peculiar volcanic structure appeared at Mont Pele in the course of the eruption of 1902, and was the subject of careful study by Professor A. Lacroix, Dr E. A. Hoovey, A. Heilprin and other observers. It appears that from fissures in the floor of the Etang Sec a viscous andesitic lava, partly quartziferous, was poured forth and rapidly solidified superficially, forming a domeshaped mass invested by a crust or carapace. According to Lacroix, the crust soon became fractured, partly by shrinkage on consolidation and partly by internal tension, and the dome grew rapidly by injection of molten matter.

Then there gradually rose from the dome a huge monolith or needle, forming a terminal spine, which in the course of its existence varied in shape and height, having been at its maximum in July 1903, when its absolute height was about 5276 ft. above sea-level. The walls of the spine, inclined at from 75° to 90° to the horizon, were apparently slickensided, or polished and scratched by friction: masses were occasionally detached and vapours were continually escaping. Several smaller needles were also formed. Some observers regarded the great spine as a solidified plug of lava from a previous outburst, expelled on a renewal of activity.

Lacroix, however, believed that it was formed by the extrusion of an enormous mass of highly viscid magma, perhaps partly solidified before emission, and he compared the formation of the dome in the crater to the structure on Santorin in 1866, described by Fouque as a " cumulo-volcano." Professor H. F. Cleland has suggested a comparison with the cone of andesite in the crater of the volcano of Toluca in Mexico, and it is said that similar formations have been observed in the volcanoes of the Andes. Dr Tempest Anderson, on visiting Pele in 1907, found a stump of the spine, consisting of a kind of volcanic agglomerate, rising from a cone of talus formed of its ruins.

The Crater

The eruptive orifice in normal volcano - the *bocca* of Italian vulcanologists - is usually situated at the bottom of a depression or cup, known as the crater. This hollow is formed and kept open by the explosive force of the elastic vapours, and when the volcano becomes dormant or extinct it may be closed, partly by rock falling from its crumbling walls and partly by the solidification of the lava which it may contain. If a renewed outburst occurs, the floor of the old crater may reopen or a new outlet may be formed at some weak point on the side of the mountain: hence a crater may, with regard to position, be either terminal or lateral.

The position of the crater will evidently be also changed on any shifting of the general axis of eruption. In shape and size the crater varies from time to time, the walls being perhaps breached or even blown away during an outburst. Hence the height of a volcanic mountain in activity, measured to the rim of the crater or the terminal peak, is not constant.

Vesuvius, for example, suffered a reduction of several hundred feet during the great eruption of 1906, the east side of the cone having lost, according to V. R. Matteucci, 120 metres.

Whilst in many cases the crater is a comparatively small circular hollow around the orifice of discharge, it forms in others a large bowllike cavity, such as is termed in some localities a "caldera". In the Sandwich Islands the craters are wide pits bounded by nearly vertical walls, showing stratified and terraced lavas and floored by a great plain of black basalt, sometimes with lakes of molten lava.

Professor W. H. Pickering compares the lava-pits of Hawaii to the crater-rings in the moon. Some of the pit-craters in the Sandwich Islands are of great size, but none comparable with the greatest of the lunar craters. Dr G. K. Gilbert, however, has suggested that the ring-shaped pits on the moon are not of volcanic origin, but are depressions formed by the impact of meteorites. Similarly the "crater" of Coon Butte, near Canyon Diablo, in Arizona, which is 4000 ft. in diameter and 500 ft. deep, has been regarded as a vast pit due to collision of a meteorite of prodigious size.

Probably the largest terrestrial volcanic crater is that of Aso-san, in the isle of Kiushiu, which is a huge oval depression estimated by some observers to have an area of at least zoo sq. m. Some of the large pit-craters have probably been formed by subsidence, the cone of a volcano having been eviscerated by extravasation of lava, and the roof of the cavity having then subsided by loss of support. The term caldera has sometimes been limited to craters formed by such collapse.

On the floor of the crater, ejected matter may accumulate as a conoidal pile; and if such action be repeated in the crater of the new cone, a succession of concentric cones will ultimately be formed. The walls of a perfect crater form a ring, giving the cone a truncated appearance, but the ring may suffer more or less destruction in the course of the history of the mountain. A familiar instance of such change is afforded by Vesuvius.

The mountain now so called, using the term in a restricted sense, is a huge composite cone built up within an old crateral hollow, the walls of which still rise as an encircling rampart on the N. and N.E. sides, and are known as Monte Somma; but the S. and S.W. sides of the ancient crater have disappeared, having been blown away during some former outburst, probably the Plinian eruption of 79. In like manner the relics of an old crater form an amphitheatre partially engirdling the Soufriere in St Vincent, and other examples of "Somma rings" are known to vulcanologists.

Much of the fragmental matter ejected from a volcano rolls down the inside of the crater, forming beds of tuff which incline towards the central axis, or have a centroclinal dip. On the contrary, the sheets of cinder and lava which form the bulk of the cone slope away from the axis, or have a dip that is sometimes described as pericentric or qua-qua-versal. According to the old "crater-of-elevation theory" held especially by A. von Humboldt, L. von Buch and Elie de Beaumont, this inclination of the beds was regarded as mainly due to upheaval. It was contended that the volcanic cone owed its shape, for the most part, to local distension of the ground, and was indeed comparable to a huge blister of the earth's crust, burst at the summit to form the "elevation crater." Palma, in the Canary Islands, was cited as a typical example of such a formation.

This view was opposed mainly by Poulett-Scrope, Sir Charles Lyell and Constant Prevost, who argued that the volcano, so far from being bladder-like, was practically a solid cone of erupted matter: hence this view came to be known as the "crater-of-eruption theory." Its general soundness has been demonstfated whenever an insight has been obtained into the internal structure of a volcano. Thus, after the eruption of Krakatoa in 1883 a magnificent natural section of the great cone of Rakata, at the S. end of the island, was exposed - the northern half having been blown away - and it was then evident that this mountain was practically a solid cone, built up of a great succession of irregular beds of tuff and lava, braced together by intersecting dykes.

The internal architecture of a volcano is rarely so well displayed as in this case, but dissections of cones, more or less distinct, are often obtained by denudation. It should be mentioned that, in connexion with the structures called *laccoliths,* there may have been an elevation, or folding, and even faulting, of the superficial rocks by subterranean intrusion of lava; but this is different from the local expansion and rupture of the ground required by the old theory. It may be noted, however, that in recent years the view of elevation, in a modified form, has not been without supporters.

Where the growth of a volcanic mound takes place from within, as in certain steep-sided trachytic cones, there may be no perceptible crater or external outlet. Again, there are many volcanoes which have no crater at the summit, because the eruptions always take place from lateral outlets. Even when a terminal pit is present, the lava may issue from the body of the mountain, and in some cases it exudes from so many vents or cracks that the volcano has been described as " sweating fire."

PARASITIC CONES

In the case of a lofty volcano the column of lava may not have sufficient ascensional force to reach the crater at the summit, or at any rate it finds easier means of egress at some weak spot, often along radial cracks, on the flanks of the mountain.

Thus at Etna, which rises to a height of more than io,800 ft., the eruptions usually proceed from lateral fissures, sometimes at least half-way down the mountain-side. When fragmental materials are ejected from a lateral vent a cinder-cone is formed, and by frequent repetition of such ejections the flanks of Etna have become dotted over with hundreds of scoria-cones much like the puys of Auvergne, the largest (Monte Minardo) rising to a height of as much as 750 ft. Hills of this character, seated on the parent mountain, are known as parasitic cones, minor cones, lateral cones.

Such subordinate cones often show a tendency to a linear arrangement, rising from vents or *bocche* along the floor of a line of fissure. Thus in 1892 a chain of five cones arose from a rift on the S. side of Etna, running in a N. and S. direction, and the hills became known as the Monti Silvestri, after Professor

Orazio Silvestri of Catania. This rift, however,was but a continuation of a fissure from which there arose in 1886 the series of cones called the Monti Gemmellaro, while this in turn was a prolongation of a rent opened in 1883. The eruption on Etna in the spring of 1910 took place along the same general direction, but at a much higher elevation. The tendency for eruptions to be renewed along old lines of weakness, which can be readily opened afresh and extended, is a feature well known to vulcanologists.

The small cones which are frequently thrown up on lava streams were admirably exemplified on Vesuvius in the eruption of 1855 and figured by J. Schmidt. The name of " driblet cones " was given by J. D. Dana to the little cones and pillars formed by jets of lava projected from blowing holes at Kilauea, the drops of lava remaining plastic and cohering as they fell. Such clots may form columns and. pyramids, with almost vertical sides. Steep-sided cones more or less of this character occur elsewhere, but are usually built up around spiracles. Small cones formed by mere dabs of lava are known trivially as " spatter cones."

FISSURE ERUPTIONS

In certain parts of the world there are vast tracts of basaltic lava with little or no evidence of cones or of pyroclastic accompaniment. To explain their formation, Baron F. von Richthofen suggested that they represent great floods of lava which were poured forth not from ordinary volcanic craters with more or less explosive violence, but from great fissures in the earth's crust, whence they may have quietly welled forth and spread as a deluge over the surface of the country. The eruptions were thus effusive rather than explosive. Such phenomena, constituting a distinct type of vulcanism, are distinguished as fissure eruptions or massive eruptions - terms which suggest the mode of extrusion and the character of the extruded matter.

As the lava in such outflows must be very fusible, it is generally of basaltic type, like that of Hawaii: indeed, the Hawaiian volcanoes, with their quiet emission of highly fluent lavas, connect the fissure eruptions with the " central eruptions," which are usually regarded as representing the normal type of activity. At the present day true fissure eruptions seem to be of rather limited occurrence, but excellent examples are furnished by Iceland. Here there are vast fields of black basalt, formed of sheets of lava which have issued from long chasms, studded in most cases with rows of small cones, but these generally so insignificant that they make no scenic features and might be readily obliterated by denudation. Dr T. Thoroddsen enumerates 87 great rifts and lines of cones in Iceland, and even the larger cones of Vesuvian type are situated on fissures.

It is believed that fissure eruptions must have played a far more important part in the history of the earth than eruptions of the familiar cone-and-crater type, the latter representing indeed only a declining phase of vulcanism. Sir Archibald Geikie, who has specially studied the subject of fissure eruptions, regards the Tertiary basaltic plateaus of N.E. Ireland and the Inner Hebrides as outflows from fissures, which may be represented by the gigantic system of dykes that form so marked a feature in the geological structure of the northern part of Britain and Ireland.

These dykes extend over an area of something like 40,000 sq. m., while the outflows form an aggregate of about 3000 ft. in thickness. In parts of Nevada, Idaho, Oregon and Washington, sheets of late Tertiary basalt from fissure eruptions occupy an area of about 200,000 sq. m., and constitute a pile at least 2000 ft. thick. In India the " Deccan traps " represent enormous masses of volcanic matter, probably of like origin but of Cretaceous date, whilst South Africa furnishes other examples of similar outflows. Professor J. W. Gregory recognized in the Kapte plains of East Africa evidence of a type of vulcanism, which he distinguished as that of " plateau eruptions. " According to him a number of vents opened at the points of intersection of lines of weakness in a high plateau, giving rise to many small cones, and the simultaneous flows of lava from these cones united to form a far-spreading sheet.

Extrusive and Intrusive Magmas

When the molten magma in the interior of the earth makes its way upwards and flows forth superficially as a stream of lava, the product is described as extrusive, effusive, effluent or eruptive; but if, failing to reach the surface, the magma solidifies in a fissure or other subterranean cavity, it is said to be intrusive or irruptive. Rocks of the former group only are sometimes recognized as strictly " volcanic, " but the term is conveniently extended, at least in certain cases, to igneous rocks of the latter type, including therefore certain hypabyssal and even plutonic rocks.

When the intrusive magma has been forced into narrow irregular crevices it forms " veins, " which may exhibit complex ramifications, especially marked in some acid rocks; but when injected into a regularly shaped fissure, more or less parallel-sided, and cutting across the planes of bedding, it forms a wall-like mass of rock termed a " dyke. " Most dykes are approximately vertical, or at least highly inclined in position. The inclination of a dyke to a vertical plane is termed its " hade. " In a cinder-cone, the lava as it rises may force its way into cracks, formed by pressure of the magma and tension of the vapours, and will thus form a system of veins and dykes, often radiating from the volcanic axis and strengthening the structure by binding the loose materials together.

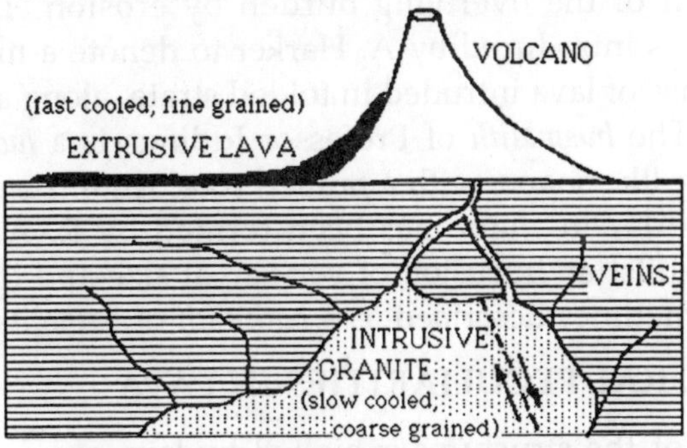

Fig. Extrusive and Intrusive Magmas

Thus, in the Valle del Bove, a huge cavity on the east side of Etna, the walls exhibit numerous vertical dykes, which by their hardness stand out as rocky ribs, forming a marked feature in the scenery of the valley. In a similar way dykes traverse the walls of the old crater of Monte Somma at Vesuvius. Exceptionally a dyke may be hollow, the lava having solidified as a crust at the margin of the fissure but having escaped from the interior while still liquid.

When molten matter is thrust between beds of tuff or between successive lava-flows or even ordinary sedimentary strata, it forms an intrusive sheet of volcanic rock known as a " sill. " A sill may sometimes be traced to its connexion with a dyke, which represents the channel up which the lava rose, but instead of reaching the surface the fluid found an easier path between the strata or perhaps along a horizontal rent. Although a dyke may represent a conduit for the ascent of lava which has flowed out superficially, yet if the lava has been removed at the surface by denudation the dyke terminates abruptly, so that its function as the former feeder of a lava-current is not evident. In other cases a dyke may end bluntly because the crack which it occupies never reached the surface.

Lava which has insinuated itself between planes of stratification may, instead of spreading out as a sheet or sill, accumulate locally as a lenticular mass, known as a *laccolith* or *laccolite.* Such a mass, in many cases rather mushroom-shaped, may force the superincumbent rocks upwards as a dome, and though at first concealed may be ultimately exposed by removal of the overlying burden by erosion. The term *phacolite* was introduced by A. Harker to denote a meniscus-shaped mass of lava intruded in folded strata, along a crest or a trough. The *bysmalith* of Professor Iddings is a *laccolith*. of rather plug-like shape, with a faulted roof. An intrusive mass quite irregular in shape has been termed by R. A. Daly a *chonolith* whilst an intrusion of very great size and illdefined form is sometimes described as a *bathylith* or *batholite.*

STRUCTURAL PECULIARITIES IN LAVA

Many of the structures exhibited by lava are due to the conditions under which solidification has been effected. A dyke, for example, may be vitreous at the margin where it has been rapidly chilled by contact with the walls of the fissure into which it was injected, whilst the main body may be lithoidal or crystalline: hence a basalt dyke will sometimes have a selvage formed of the basaltic glass known as tachylyte. A similar glass may form a thin crust on certain lava-flows. In a homogeneous vitreous lava, contraction on solidification may develop curved fissures, well seen in the delicate " perlitic " cracks of certain obsidians, indicating a tendency to assume a globular structure. This structure becomes very distinct by the development of " spherulites, " or globular masses with a radiating fibrous structure, sometimes well seen in devitrified glass. Occasionally the spherulitic bodies in lava are hollow, when they are known as lithophyses, of which excellent examples occur at Obsidian Cliff in the Yellowstone National Park, as described by Professor Iddings. Globular structure on a large scale is sometimes displayed by lavas, expecially those of basic type, such as the basalt of Aci Castello in Sicily, which was probably formed, according to Professor Gaetano Platania, by flow of the lava into submarine silt, relics of which still occur between the spheroids.

Joints, or cracks formed by shrinkage on solidification, may divide a sheet of lava into columns, as familiarly seen in basalt, where the rock often consists of a close mass of regular polygonal prisms, mostly hexagonal. Each prism is divided at intervals by transverse joints, more or less curved, so that the portions are united by a slight ball-and-socket articulation. As the long axes of the columns lie at right angles to the cooling surface they are vertical in a horizontal sheet of lava, horizontal in a vertical dyke, and inclined or curved in other cases. It sometimes happens that in a basaltic dyke the formation of the prisms, having started from the opposite walls as chilling surfaces, has not been completed; and hence the prisms fail to meet in the middle.

A spheroidal structure is often developed in basalt columns by weathering, the rock exfoliating in spherical shells, rather like the skins of an onion: such a structure is characteristically shown at the Kasekellar, known also as the Elfen Grotto, at Bertrich, near Alf on the Mosel, where the pillars of the lava are broken into short segments which suggest by their flattened globular shape a pile of Dutch cheeses. Although prismatic jointing, or columnar structure, is most common in basalt, it occurs also in other volcanic rocks. Fine columns of obsidian, for instance, are seen at Obsidian Cliff in the Yellowstone Park, where the pillars may be 50 ft. or more in height. Such an occurrence, however, is exceptional. Vitreous lavas often show fluxion structure in the form of streaks, bands or trains of incipient crystals, indicating the flow of the mass when viscous. The character of this structure is related to the viscosity of the lava. Those structural peculiarities which depend mainly on the presence of vapour, such as vesiculation, have been already noticed, and the porphyritic structure has likewise been described.

Submarine Volcanoes

Considering how large a proportion of the face of the earth is covered by the sea, it seems likely that 'volcanic eruptions must frequently occur on the ocean-floor. When, as occasionally though not often happens, the effects of a submarine eruption are observed during the disturbance, it is seen that the surface of the sea is violently agitated, with copious discharge of steam; the water passes into a state of ebullition, perhaps throwing up huge fountains; shoals of dead fishes, with volcanic cinders, bombs and fragments of pumice, float around the centre of eruption, and ultimately a little island may appear above sea-level. This new land is the peak of a volcanic cone which is based on the sea-floor, and if in deep water the submarine mountain must evidently be of great magnitude. Christmas Island in the Indian Ocean, described by Dr C. W. Andrews, appears to be a volcanic mountain, with Tertiary limestones,. standing in water more than 14,000 ft. deep. Many volcanic islands, such as those abundantly scattered over the Pacific, must have started as submarine volcanoes which reached the surface either by continued upward growth or by upheaval of the sea-bottom. Etna began its long geological history by submarine eruptions in a bay of the Mediterranean, and Vesuvius in like manner represents what was originally a volcano on the sea-floor. As the ejectamenta from a submarine vent accumulate on the sea-bottom they become intermingled with relics of marine organisms, and thus form fossiliferous volcanic tuffs. By the distribution of the ashes over the sea-floor, through the agency of waves and currents, these tuffs may pass insensibly into submarine deposits of normal sedimentary type.

Fig. Submarine Volcanoes

One of the best examples of a submarine eruption resulting in the formation of a temporary island occurred in 1831 in the Mediterranean between Sicily and the coast of Africa, where the water was known to have previously had a depth of too fathoms. After the usual manifestations of volcanic activity an accumulation of black cinders and ashes formed an island which reached at one point a height of 200 ft., so that the pile of erupted matter had a thickness of about 800 ft. The new island, which was studied by Constant Prevost, became known in England as Graham's Island, in France as Ile Julie and in Italy by various names as Isola Ferdinandea. Being merely a loose pile of scoriae, it rapidly suffered erosion by the sea, and in about three months was reduced to a shoal called Graham's Reef.

In 1891 a submarine eruption occurred near the isle of Pantellaria in the same waters, and the eruptive centre was termed by Professor H. S. Washington and Foerstner volcano, but it gave rise to no island. A well-known instance of a temporary volcanic island was furnished by Sabrina - an islet of cinders thrown up by submarine eruptions in 1811, off the coast of St Michael's, one of the Azores. The island of Bogosloff, or Castle island, in Bering Sea, about 40 m. W. of Unalaska Island, is a volcanic mass which was first observed in 1796 after an eruption. In 1883 another eruption in the neighbouring water threw up a new volcanic cone of black sand and ashes, known as New Bogosloff or Fire Island, situated about half a mile to the N.W. of Old Bogosloff, with which it was connected by a low beach. Another island, called Perry Island, larger than either of the others, made its appearance in the neighbourhood about the time of the great earthquake in California in 1906. It is reported that some of these islands have since disappeared.

Mud Volcanoes

Mud volcanoes are small conical hills of clay which discharge, more or less persistently, streams of fine mud, sometimes associated with naphtha or petroleum, and usually with bubbles of gas. As the mud is generally saline, the hills are known also as " salses." The gases are chiefly hydrocarbons, often with more or less sulphuretted hydrogen and carbon dioxide, and sometimes with nitrogen. Though generally less than a yard in height, the cones may in exceptional cases rise to an elevation of as much as Soo ft. The mud oozes from the top and spreads over the sides, or is spurted forth with the gases. Occasionally the discharge is vigorous, mud and stones being thrown up to a considerable height, sometimes accompanied by flames due to combustion of the hydrocarbons.

Fig. Mud of Volcanoes

Mud volcanoes occur in groups, and have a wide distribution. They are known in Iceland; in Modena; at Taman and Kertch, in the Crimea; at Baku on the Caspian; in Java and in Trinidad: Humboldt described those near Turbaco, in Colombia. In Sicily they occur near Girgenti, and a group is known at Paterno on Etna. Bythe Sicilians they are termed, *maccalube,* a word of Arabic origin. The " paint-pots " of the Yellowstone National Park are small mud volcanoes.

Many, so-called mud volcanoes appear to be due to the derangement of subterranean water-flow or to landslips in connexion with earthquakes, whilst others may be referable to certain chemical reactions going on underground; but there are others again which seem to be truly of volcanic origin. Hot water and steam escaping through clays, or crumbling tuffs reduced to a clayey condition, may form conical mounds of pasty material, through which mud oozes and water escapes.

Geysers are closely related to volcanoes, but in consequence of their special interest they are treated separately. For natural steam-holes and other phenomena connected with declining vulcanicity.

GEOGRAPHICAL DISTRIBUTION OF VOLCANOES

It is matter of frequent observation that volcanoes are most abundant in regions marked by great seismic activity. Although the volcano and the earthquake are not usually connected in the direct relation of cause and effect, yet in many cases they seem referable to a common origin. Both volcanic extrusion and crustal movement may be the means of relieving local strains in the earth's crust, and both are found to occur, as might reasonably be expected, in many parts of the earth where folding and fracture of the rocks have frequently happened and where mountain-making appears to be still in progress.

Thus, volcanoes may often be traced along zones of crustal deformation, or folded mountain-chains, especially where they run near the borders of the oceanic basins. They are frequently associated with the Pacific type of coast-line.

The most conspicuous example of linear distribution is

furnished by the great belt of volcanoes, coinciding for the most part with a band of seismic disturbance, which engirdles intermittently the huge basin of the Pacific; though here, as elsewhere in studying volcanic topography, regard must be paid to dormant and extinct centres as well as to those that are active at the present time. As volcanoes are in many cases ranged along what are commonly regarded as lines of fracture, it is not surprising that the centres of most intense vulcanicity are in many cases situated at the intersection of two or more fracture-lines. On the eastern side of the Pacific Ocean the great volcanic ring may be traced, though with many and extensive interruptions, from Cape Horn to Alaska.

In South America the chain of the Andes between Corcovado in the south and Tolima in the north is studded at irregular intervals with volcanoes, some recent and many more extinct, including the loftiest volcanic mountains. in the world. The linear arrangement, often a marked feature in the distribution of volcanoes, is well exemplified in the general north-andsouth trend of the Andean ranges, the volcanoes being situated along the orographic axis.

These folded mountains with their volcanoes also illustrate the close relationship to the sea so frequently observed in volcanic topography, a relationship, however, not without many exceptions. The volcanic rock called andesite was so named by L. von Buch from its characteristic occurrence in the Andes. It is notable that the volcanic rocks throughout the great Pacific belt present much similarity in composition. The volcanoes of Ecuador have been described in detail by A. Stiibel and others.

Central America contains a large number of active volcanoes and solfataras, many of which are located in the mountains parallel to the western coast. Conseguina, on the south side of the Gulf of Fonseca, is remarkable for its eruption in 1835, when an enormous, volume of ash was ejected and the summit of the mountain blown away. Izalco, in San Salvador, came into existence in 1770, and is habitually active. In the centre of Lake Ilopango in Salvador,. which possibly occupies an ancient crater, a volcanic island arose in 1880 and attained a height of 160 ft. Guatemala is peculiarly rich in volcanoes, as described by Dr Tempest Anderson, who visited the country in 1907.

The Cerro Quemado, or the Volcano of Quezaltenango, was the scene of a great eruption in 1785. At the Volcano of Santa Maria there was an outburst in 1902 more violent than the simultaneous eruptions in the Lesser Antilles. The cones of Guatemala include the Volcan de Fuego and the Volcan de Agua, the former often active in historic times, whilst the latter is notable for the flood which in 1541 swept down from the mountain and destroyed Old Guatemala, but this flood was probably not of volcanic origin.

The plateau of Mexico is the seat of several active volcanoes which occur in a band stretching across the country from Colima in the west to Tuxtla near Vera Cruz. The highest of these volcanic mountains is Orizaba, or Cithaltepetl, rising to an altitude of 18,200 feet, and known to have been active in the 16th century. Popocatepetl ("the smoking mountain") reaches a height of about, 17,880 ft., and from its crater sulphur was at one time systematically collected. It came into existence rapidly during an eruption which began in September 1759, when it was said by unscientific observers that the ground became suddenly inflated from below.

The cone, though not of exceptional magnitude, is situated in an elevated district, and its summit rises to about 4330 ft. above sea-level. In the neighbourhood of Jorullo there are three subordinate cones of similar character known as *volcancitos*, with great numbers of small mounds of cinder and ash formed around fumaroles on the lava, and locally called *hornitos*, or " little ovens. " The streams of basaltic lava from Jorullo form rough barren surfaces, which pass under the name of *malpays*, or bad lands.

In the United States very few volcanoes are active at the present day, though many have become extinct only in times that are geologically recent. An eruption occurred in 1857 at Tres Virgines, in the south of California, and the cinder cone on Lassen's Peak was also active in the middle of the 19th century. The Mono Valley craters and Mount Shasta, in

California, are extinct. The Cascade Range contains numerous volcanic peaks, but only few show signs of activity. 1Vlount Hood, in Oregon, exhales vapour, as also does Mount Rainier in Washington. Mount St Helens was in eruption in 1841 and 1842; and Mount Baker the most northern of the volcanoes connected with the Cascade Range, is said to have been active in 1843. Few volcanic peaks occur in the Rocky Mountains, but evidence of lingering activity is very marked in the geysers and hot springs of the Yellowstone National Park. The earth's internal heat is also manifested at many points elsewhere, as at Steamboat Springs on the Virginia Range, an offshoot of the Sierra Nevada, and in the Comstock Lode.

Volcanic activity is prominent in Alaska, along the Coast Range and in the neighbouring islands. The crater of Mount Edgecumbe, in Lazarus Island, is said to have been active in 1796, but this is doubtful. Mount Fairweather has probably been in recent activity, and the lofty cone of Mount Wrangell, on Copper river, is reported to have been in eruption in 1819. In the neighbourhood of Cook's Inlet there are several volcanoes, including the island of St Augustine. Unimak Island has two volcanoes, which have supplied the natives, with sulphur and obsidian; one of these volcanoes being Mount Shishaldin, a cone rivalling Fusiyama in graceful contour. The.

Aleutian volcanic belt is a narrow, curved chain of islands, extending from Cook's Inlet westwards for nearly 1600 m. It is notable that the convexity of the curve faces the great ocean, as has been observed in other cases, the arcs following the direction of the rock-folds. Açcording to Professor I. C. Russell, an authority on the volcanoes of N. America, there are in the Aleutian Islands and in the peninsula no fewer than 57 craters, either active or recently extinct.

From the Aleutian Islands the volcanic band of the Pacific changes its direction, and passing to the peninsula of Kamschatka, where 14 volcanoes are said to be active, turns southwards and forms the festoon of the Kurile Islands. Here again the convexity of the insular arc is directed towards the ocean. This volcanic archipelago leads on to the great islands of Japan, where the volcanoes have been studied by Professor

J. Milne, who also described those of the Kuriles. Of the 54 volcanoes recognized as now active or only recently extinct in Japan, the best known is the graceful cone of the sacred mountain Fusiyama, but others less pretentious are far more dangerous. The great eruption of Bandaisan, about 120 m. N. of Tokio, which occurred in 1888, blew off one side of the peak called Kobandai, removing, according to Professor Sekiya's estimate, about 2982 million tons of material. Aso-san in Kiushui, the southernmost large island of Japan, is notable for the enormous size of its crater. In the Bonin group of islands volcanic activity is indicated by such names as Volcano Island and Sulphur Island.

South of the Japanese archipelago the train of volcanoes passes through some small islands in or near the Loo Choo group and thence onwards by Formosa to the Philippine Islands, where subterranean activity finds abundant expression in earthquakes and volcanoes. After leaving this region the linear arrangement of the eruptive centres becomes less distinctly marked, for almost every island in the Moluccas and the Sunda Archipelago teems with volcanoes, solfataras and hot springs. Possibly, however, a broken zone may be traced from the Moluccas through New Guinea and thence to New Zealand, perhaps through eastern Australia (for though no active volcanoes are known there, relics of comparatively recent activity are abundant); or again by way of the Bismarck Archipelago, the Solomon Islands, the New Hebrides, the Fiji Islands and Kermodoc Island.

The great volcanic district in New Zealand is situated in the northern part of North Island, memorable for the eruption of Tarawera in 1886. This three-peaked mountain on the south side of Lake Tarawera, not previously known to have been active, suddenly burst into action; a huge rift opened, and Lake Rotomahana subsided, with destruction of the famous sinter terraces. The crater of Tongariro is in the solfatara stage, whilst Mount Ruapehu is regarded as extinct. On White Island in the Bay of Plenty the cone of Wharkari is feebly active.

Far to the south, on Ross Island, off South Victoria Land, in Antarctica, are the volcanoes of Erebus and Terror, the former of which is active. These are often regarded as remotely related to the Pacific zone, but Dr G. T. Prior has shown that the Antarctic volcanic rocks which he examined belonged to the Atlantic and not the Pacific type.

Within the great basin of the Pacific, imperfectly surrounded by its broken girdle of volcanoes, there is a vast number of scattered islands and groups of islands of volcanic origin, rising from deep water, and hiving in many cases active craters. The most important group is the Hawaiian Archipelago, where there is a chain of at least fifteen large volcanic mountains - all extinct, however, with the exception of three in Hawaii, namely Mauna Loa, Kilauea and Hualalai; and of these Hualalai has been dormant since 1811.

It is notable that the two present gigantic centres of activity, though within 20 m. of each other, appear to be independent in their eruptivity. Several of the Hawaiian Islands, as pointed out by J. D. Dana, who was a very high authority on this group, consist of two volcanoes united at the base, forming volcanic twins or doublets.

The volcanic regions of the Pacific are connected with those of the Indian Ocean by a grand train of islands rich in volcanoes, stretching from the west of New Guinea through the Moluccas and the Sunda Islands, where they form a band extending axially through Java and Sumatra. Here is situated the principal theatre of terrestrial vulcanicity, apparently representing an enormous fissure, or system of fissures, in the earth's crust, sweeping in a bold curve, with its convexity towards the Indian Ocean.

Numerous volcanic peaks occur in the string of small islands to the east of Java - notably in Flores, Sumbawa, Lombok and Bali; and one of the most terrific eruptions on record in any part of the world occurred in the province of Tomboro, in the island of Sumbawa, in the year 1815. Java contains within its small area as many as 49 great volcanic mountains - active, dormant and extinct. The largest is Smerin, about 12,000 ft. high, but the most regularly active is said to be Gownong Lamongang, which is in almost uninterrupted activity, emitting usually only ashes and vapour, though in

1883 lava streamed forth. Many of the Javanese volcanoes present marked regularity of contour, with the sides of the cones rather symmetrically furrowed by tropical rains and probably ridged by ash-slides.

The radial furrows on volcanic cones are sometimes known as " barrancos." The little uninhabited island of Krakatoa in the Strait of Sunda appears to be situated at a volcanic node, or the intersection of two curved fissures, and it is believed that the island itself represents part of the basal wreck of what was once a volcano of gigantic size. After two centuries of repose, a violent catastrophe occurred in 1883, whereby the greater part of the island was blown away. This eruption and its effects were made the subject of careful study by Verbeek, B g on and Judd.

Through the great island of Sumatra, a chain of volcanoes runs longitudinally, and may possibly be continued northwards in the Bay of Bengal by Barren Island and Norcondam - the former an active and the latter an extinct volcano. On the western side of the Indian Ocean a small volcanic band may be traced in the islands of the Mascarene group, several craters in Reunion being still active. Far south in the Indian Ocean are the volcanic islands of New Amsterdam and St Paul.

The Comoro Islands in the channel of Mozambique exhibit volcanic activity, whilst in East and Central Africa there are several centres, mostly extinct but some partially active, associated with the Rift Valleys. The enormous cones of Kenia and Kilimanjaroo are extinct, but on Kibo, one of the summits of the latter, a crater is still preserved. The Mfumbiro volcanoes, S. of Lake Edward, rise to a height of more than 14,700 feet. Kirunga, N. of Lake Kivu, is still partially active. Elgon is an old volcanic peak, but Ruwenzori is not of volcanic origin. On the west side of Africa, the Cameroon Peak is a volcano which was active in 1909, and the island of Fernando Po is also volcanic. Along the Red Sea there are not wanting several examples of volcanoes, such as Jebel Teir. Aden is situated in an old crater.

Passing to the Atlantic, a broken band of volcanoes, recent

and extinct, may be traced longitudinally through certain islands, some of which rise from the great submarine ridge that divides the ocean, in part of its length, into an eastern and a western trough. The northern extremity of the series is found in Jan Mayen, an island in the Arctic Ocean, where an eruption occurred in 1818. Iceland, however, with its wealth of volcanoes and geysers, is the most important of all the Atlantic centres.

According to Dr T. Thoroddsen there are in Iceland about 130 post-glacial volcanoes, and it is known that from 25 to 30 have been in eruption during the historic period. Many of the Icelandic lava-flows, such as the immense flood from Laki in 1783, are referable to fissure eruptions, which are the characteristic though not the exclusive form of activity in this island. Probably this type was also responsible for the sheets of old lava in the terraced hills of the Faroe Islands, to which may have been related the Tertiary volcanoes of the west of Scotland and the north of Ireland.

An immense gap separates the old volcanic area of Britain from the volcanic archipelagoes of the Azores, the Canaries and the Cape Verd Islands. Palma - a little island in the Canary group, with a caldera or large crater at its summit, from which fissures or barrancos radiate - is famous in the history of vulcanology, in that it furnished L. von Buch with evidence on which he founded the " crater-ofelevation " theory. The remaining volcanic islands of the Atlantic chain, all now cold and silent, include Ascension, St Helena and Tristan da Cunha, whilst in the western part of the South Atlantic are the small volcanic isles of Trinidad and Ferdinando do Noronha. St Paul's rocks appear also to be of volcanic origin.

One of the most important volcanic regions of the world is found in the West Indies, where the Lesser Antilles - the scene of the great catastrophes of 1902 - form a string of islands, stretching in a regular arc that sweeps in a N. and S. direction across the eastern end of the Caribbean Sea. Subject to frequent seismic disturbance, and rich in volcanoes, solfataras and hot springs, these islands seem to form the summit of a great earth-fold which, rising as a curved ridge from deep water, separates the Caribbean Sea from the Atlantic. The volcanoes are situated on the inner border of the curve. It is notable that the Antilles and the Sunda Islands, two of the grandest theatres of vulcanicity on the face of the earth, are situated at the antipodes of each other - one being apparently an eastern and the other a western offshoot of the great Pacific girdle.

The European volcanoes, recent and extinct, may be regarded as representing rather ill-defined branches thrown off eastwards from the Atlantic band. Vesuvius is the only active volcano on the mainland, but in the Mediterranean there are Etna on the coast of Sicily; the Lipari Islands, with Stromboli and Vulcano in chronic activity; and farther to the east the archipelago of Santorin, where new islands have appeared in historic times. Submarine eruptions have occurred also between Sicily and the coast of Africa; one in 1831 having given rise temporarily to Graham's Island, and another in 1891 appearing near Pantellaria, itself a volcanic isle.

Of the extinct European volcanoes, some of the best known are in Auvergne, in the Eifel, in Bohemia and in Catalonia, whilst the volcanic land of Italy includes the Euganean hills, the Alban hills, the Phlegraean Fields, &c. The great lakes of Bolsena and Bracciano occupy old craters, and many smaller sheets of water are on similar sites. The volcanic islands no longer active include Ischia, with the great cone of Epomeo which was in a state of eruption in 1301; the Ponza Islands, Nisida, Vivera and others near Naples; and several in the Greek archipelago, such as Milos, Kimolos and Polinos.

From the eastern end of the Mediterranean evidence of former volcanic activity may be traced into Asia Minor and thence to Armenia and the Caucasus. East of Smyrna there is a great desolate tract which the ancients recognized as volcanic and termed the Catacecaumene. The volcanic districts of Lydia were studied by Professor H. S. Washington.

In the plateau of Armenia there are several extinct volcanic mountains, more or less destroyed, of which the best known is Ararat. Nimrud Dagh on the shore of Lake Van is said to have been in eruption in the year 1441. Dr F. Oswald has described the volcanoes of Armenia. Of the volcanoes in

Persian territory not now active, Demavend, south of the Caspian, is an important example. Elburz is also described as an old volcano. It has been said that in Central Asia there are certain vents still active, and recent volcanic rocks are known from the Przhevalsky chain and other localities.

The number of volcanoes known to be actually active on the earth is generally estimated at between 300 and 400, but there is reason to believe that this estimate is far too low. If account be taken of those volcanic cones which have not been active in historic time, the total will probably rise to several thousands. The distribution of volcanoes at various periods of the earth's history, as revealed by the local occurrence of volcanic rocks at different horizons in the crust of the earth. Periods of great earthmovement have been marked by exceptional volcanic activity.

Causes of Vulcanicity

The cause of vulcanicity two problems demand attention: first the origin of the heat necessary for the manifestation of volcanic phenomena, and secondly the nature of the force by which the heated matter is raised to the surface and ejected. According to the old view, which assumed that the earth was a spheroid of molten matter invested by a comparatively thin crust of solid rock, the explanation of the phenomena appeared fairly simple. The molten interior supplied the heated matter, while the shrinkage of the cooling crust produced fractures that formed the volcanic channels through which it was assumed the magma might be squeezed out in the process of contraction.

When physicists urged the necessity of assuming that the globe was practically solid, vulcanologists were constrained to modify their views. Following a suggestion of W. Hopkins of Cambridge, they supposed that the magma, instead of existing in a general central cavity, was located in comparatively small subterranean lakes. Some authorities again, like the Rev. O. Fisher, regarded the magma as constituting a liquid zone, intermediate between a solid core and a solid shell.

If solidification of the primitive molten globe proceeded from the centre outwards, so as to form a sphere practically solid, it is conceivable that portions of the original magma might nevertheless be retained in cavities, and thus form " residual lakes." Although the mass might be for the most part solid, the outer portion, or " crust," could conceivably have a honeycombed structure, and any magma retained in the cells might serve indirectly to feed the volcanoes. Neighbouring volcanoes seem in some cases to draw their supply of lava from independent sources, favouring the idea of local cisterns or "intercrustal reservoirs."

It is probable, however, that subterranean reservoirs of magma, if they exist, do not represent relics of an original fluid condition of the earth, but the molten material may be merely rock which has become fused locally by a temporary development of heat or more likely, by a relief of pressure. It should be noted that the quantity of magma required to supply the most copious lava-flows is comparatively small, the greatest recorded outflow ot having exceeded, it is said, six cubic miles; and even this estimate is probably too high. Whilst in many cases the magma-cisterns may be comparatively small and temporary, it must be remembered that there are regions where the volcanic rocks are so similar throughout as to suggest a common origin, thus needing intercrustal reservoirs of great extent and capacity. It has been suggested that comparatively small basins, feeding individual volcanoes, may draw their supply from more extensive reservoirs at greater depths. Much speculation has been rife as to the source of the heat required for the local melting of rock. Chemical action has naturally been suggested, especially that of superficial water, but its adequacy may be doubted. After Sir Humphry Davy's dis covery of the metals of the alkalis, he thought that their remarkable behaviour with water might explain the origin of subterranean heat; and in more recent years others have seen a local source of heat in the oxidation of large deposits of iron, such as that brought up in the basalt of Disco Island in Greenland. It has been assumed by Moissan and by Gautier that water might attack certain metallic carbides, if they occur as subterranean deposits, and give rise to some of the products characteristic of volcanoes. But it seems that all such action must be very limited, and utterly inadequate to the general explanation of volcanic phenomena.

At the same time it must be remembered that access of water to a rock already heated may have an important. physical effect by reducing its melting point, and may thus greatly assist in the production of a supply of molten matter. The admission of surface-waters to heated rocks is naturally regarded as an important source of motive power in consequence of the sudden generation of vapour, but it is doubtful to what extent it may contribute, if at all, to the origin of volcanic heat.

According to Robert Mallet a competent source of subterranean heat for volcanic phenomena might be derived from the transformation of the mechanical work of compressing and crushing parts of the crust of the earth as a consequence of secular contraction. This view he worked out with much ingenuity, supporting it by mathematical reasoning and an appeal to experimental evidence. It was claimed for the theory that it explained the linear distribution of volcanoes, their relation to mountain chains, the shallow depth of the foci and the intermittence of eruptive activity.

A grave objection,. however, is the difficulty of conceiving that the heat, whether due to crushing or compression, could be concentrated locally so as to produce a sufficient elevation of temperature for melting the rocks. According to the calculations of Rev. O. Fisher, the crushing could not, under the most favourable circumstances, evolve heat enough to account for volcanic phenomena.

Since pressure raises the melting-point of any solid that expands on liquefaction, it has been conjectured that many deep-seated rocks, though actually solid, may be potentially liquid; that is, they are maintained in a solid state by pressure: only. Any local relief of pressure, such as might occur in the folding and faulting of rocks, would tend, without further accession of heat, to induce fusion. But although moderate pressure raises the fusing-point of most solids, it is believed,, from modern researches, that very great pressures may have a contrary effect.

It is held by Professor S. Arrhenius that at great depths in. the earth the molten rock, being above its critical point, can. exist only in the gaseous condition; but a gas under enormous pressure may behave, so far as compressibility is concerned, like a rigid solid. He concludes, from the high density of the earth as a whole and from other considerations, that the central part of our planet consists of gaseous iron (about 80% of the earth's diameter) followed by a zone of rock magma in a gaseous condition (about 15%), which passes insensibly outwards into liquid rock (4%), covered by a thin solid crust (less than i % of diameter).

If water from the crust penetrates by osmosis through the sea-floor to the molten interior, it acts, at the high temperature, as an acid, and decomposes the silicates of the magma. The liquid rock, expanded and rendered more mobile by this water, rises in fissures, but in its ascent suffers cooling, so that the water then loses its power as an acid and is displaced by silicic acid, when the escaping steam gives rise to the explosive phenomena of the volcano.

The earth was formed by the accretion of vast numbers of small cosmical bodies called planetesimals, the original heat of the earth's interior was due chiefly to the compression of the growing globe by its own gravity. The heat, proceeding from the centre outwards, caused local fusion of the rocks, though without forming distinct reservoirs of molten magma, and the fused matter charged with gases rose in liquid threads or tongues, which worked their way upwards, some reaching the superficial part of the earth and escaping through fissures in the zone of fracture, thus giving rise to volcanic phenomena. It is held that the explosive activity of a volcano is due to the presence of gases which have been brought up from the interior of the earth, whilst only a small and perhaps insignificant part is played by water of superficial origin.

Entirely new views of the origin of the earth's internal heat have resulted from the discovery of radioactivity. An ingenious hypothesis was enunciated by Major C. E. Dutton, who found in the radioactivity of the rocks a sufficient source of heat for the explanation of all volcanic phenomena. He believes that the development of heat arising from radioactivity may gradually bring about the local melting of the rocks so as to form large subterranean pools of magma, from which the volcanoes may be supplied.

The supply is usually drawn from shallow sources, probably, according to Dutton, from a depth of not more than three or rarely four miles, and in some cases at not more than a mile from the surface.

If the water in the local magma should attain sufficient expansive power, it will rupture the overlying rocks and thus give rise to a volcanic eruption. When the reservoir becomes exhausted the eruption ceases, but if more heat be generated by continued radioactivity further fusion may ensue, and in time the eruption be repeated. According, however, to Professor Joly, it is improbable that sufficient heat for the manifestation of volcanic phenomena could be developed by the local radioactivity of the rocks in the upper part of the earth's crust.

Chapter 3

Ground Water

Ground Water is the water that saturates the pores and cracks in soil and rock beneath the land surface. It comes to the surface in springs, and also swells the volume of many streams by seeping into them from their beds and banks. Ground water is also obtained for human use by sinking wells.

SOURCE OF GROUND WATER

Water at the earth's surface has one obvious source—rain and snow—and, as Perrault showed long ago there are good reasons for considering these sources as practically the only ones. They are also the ultimate source of practically all ground water. Most soils and rocks are porous. Small voids and openings occur between grains of soil, larger openings are formed by burrowing animals, and cracks are formed by the shrinkage of drying soil clays.

Even well-consolidated rocks may be riven by joint cracks or other openings into which water can penetrate. After a rain part of the water seeps into these openings and percolates beneath the land surface. Some of the rain water that enters the soil remains near the surface, absorbed by the soil colloids, or held against gravity in the smaller voids by surface tension (*capillarity,* the force that pulls water up a very slender tube and holds it there against the action of gravity).

Deeper water may even be drawn back toward the surface of the ground by capillarity, thus helping to replace the losses from evaporation and transpiration by plants. At least a part of the rain water, however, percolates deeper and deeper through the larger soil openings until it ultimately reaches a zone beneath the surface where all the pores in the rock are completely filled with water.

THE WATER TABLE

The name *water table* is given to the upper surface of this zone below which all pores are filled with water. Except for the discontinuous films and irregular masses of water held by capillarity, or in process of transit downward, the pores above the water table are filled with air, and so the zone between the water table and the surface is called the *zone of aeration.* Here weathering and other chemical changes go on under oxidizing conditions. Below the water table is the *zone of saturation.*

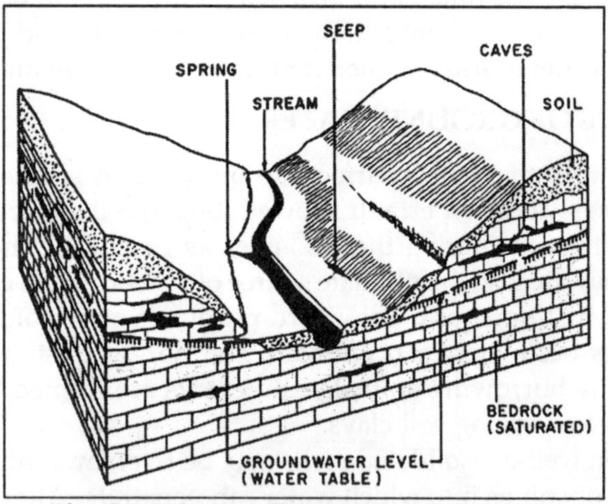

Fig. Cross Section Showing the Water Table

In most places the water table is a few feet, or a few tens of feet, below the ground surface, but in arid regions it may be hundreds of feet below. In marshes, on the other hand, the water table practically coincides with the ground surface, as it does at the edge of surface-water bodies such as lakes and streams. If many water wells have been dug, the form of the water table below the ground surface can be readily determined and accurate contour maps of it drawn from the elevation of the water surface in the various wells. Such maps show that, in general, the water table is a subdued replica of the surface topography. It rises under the hills and sinks beneath the valleys, but is usually much smoother than the land surface. The level of the water table also fluctuates slightly with the seasons.

SUBSURFACE PORES, CRACKS, AND CHANNELS

If subsurface pores are abundant and large below the water table, water in considerable volume can be pumped from wells. When wells are pumped, water from pores in the nearby rock flows rapidly into the well and replenishes the water being extracted. On the other hand, if the pores are small and widely spaced, only a little water can be recovered by pumping because replenishment is slow. The shapes, sizes, and aggregate volume of the open spaces—pore spaces—in each kind of soil and rock are thus primary factors in determining the occurrence, amount, and rate of movement of ground water.

Porosity and Permeability

Porosity is the ratio of pore volume to total volume, expressed as percentage. It is generally determined by testing samples of rock in the laboratory. Porosities of most uncemented elastic sediments range between 12 and 45 per cent. Variations in the *shapes* of the grains, their *sorting,* their *packing,* and the *degree* of cementation cause these differences.

Mineral and rock grains vary in shape from thin plates and irregular angular bodies to nearly perfect spheres. The arrangement of the grains with respect to one another (the kind of *packing*) greatly affects the amount of pore space. Commonly, the packing is loose when a sediment is first deposited, and the pore space is relatively high. This space is then progressively reduced by *compaction* from the pressure of new sediments subsequently deposited above. *Cementation* (deposition of mineral matter in the pores) further reduces the pore space. Closely packed spheres of uniform size have 26 per cent pore space. This is true whether the spheres are 1 millimeter in diameter, 5 feet in diameter, or any other size. Porosities above 26 per cent usually indicate nonuniform packing, or that some grains are themselves porous. Shape is also a factor, but the presence of nonspherical grains may either raise or lower the porosity. Porosities below 26 per cent usually result from poor sorting, or from compaction and cementation of the voids between grains angular bodies to nearly perfect spheres.

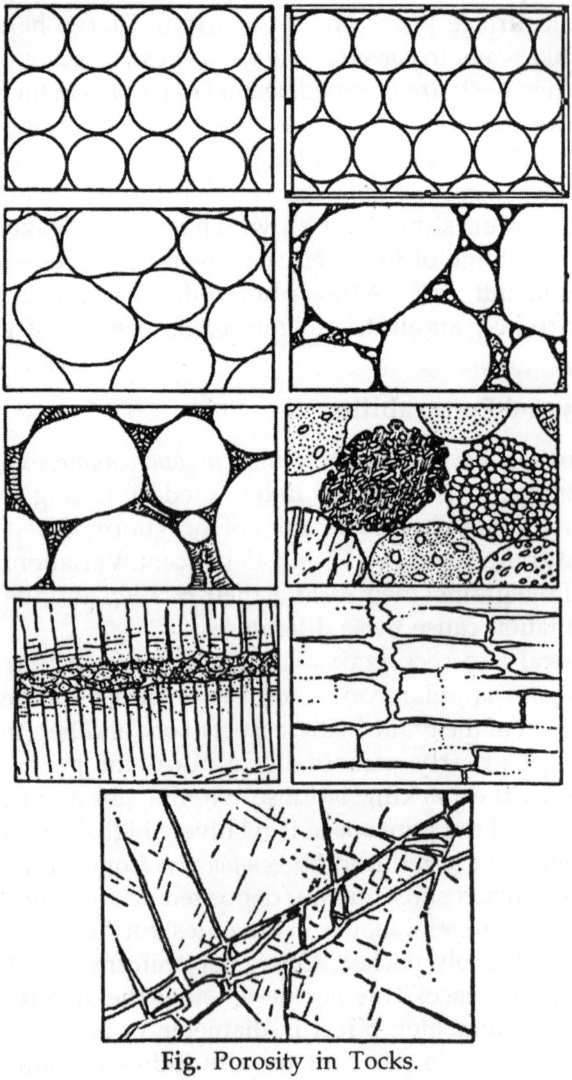

Fig. Porosity in Tocks.

The arrangement of the grains with respect to one another (the kind of *packing*) greatly affects the amount of pore space. Commonly, the packing is loose when a sediment is first deposited, and the pore space is relatively high. This space is then progressively reduced by *compaction* from the pressure of new sediments subsequently deposited above. *Cementation* (deposition of mineral matter in the pores) further reduces the pore space.

Closely packed spheres of uniform size have 26 per cent pore space. This is true whether the spheres are 1 millimeter in diameter, 5 feet in diameter, or any other size. Porosities above 26 per cent usually indicate nonuniform packing, or that some grains are themselves porous. Shape is also a factor, but the presence of nonspherical grains may either raise or lower the porosity. Porosities below 26 per cent usually result from poor sorting, or from compaction and cementation of the voids between grains.

The total capacity of rock or soil to *hold* water is determined by the porosity, but the capacity to *yield* water to the pump depends on the size of the pores as well as on the total amount of pore space. Not all the water in the pores will flow toward a pumping well. Much is retained as water films stuck to the walls of the pores. In rocks with very small pores practically all the water may thus be retained, even though the porosity is high. *Permeability, the capacity of a porous medium to transmit a liquid,* and not porosity, is, therefore, the most important physical property to be determined in estimating the reservoir capacity of a water-bearing material.

Permeability depends more on the size of the pore openings than on the percentage of pore space. A gravel with 20 per cent of pore space is much more permeable to ground water than a clay with 35 per cent porosity.

Laboratory measurements of permeability, as well as the distribution and yield of springs and wells, indicate that there is a fairly definite minimum size of pores—about 0.05 millimeters—through which water will move freely. Sediments with smaller pores, such as clay, silt, and shale, are impermeable even though their porosity may be high. Larger pores are present in sand and gravel and also in many incompletely cemented sandstones, conglomerates, jointed lava flows, and porous limestones.

AQUIFERS

A body of rock or loose surface material that is permeable as well as porous, and so can yield water to wells, is called an *aquifer*. Some aquifers may be empty of water, at least temporarily. Even in an aquifer, water movement is extremely slow, usually a fraction of a foot or a few feet per day, except in the immediate vicinity of a well or spring.

Most aquifers are sheets of sand or gravel, or beds of sandstone, limestone, or other permeable rock. A few aquifers are narrow sinuous bodies of gravel that fill former stream courses, but these are by no means as common as the popular expression "underground stream" indicates.

Water Witching

Before the Nineteenth Century men thought that ground water flowed in definite underground streams just as surface water does. They reasoned that when one dug a well, if he were lucky, his well intersected one of these streams and produced a fine flow of water. A dry well, or one that produced only a little water, supposedly had failed to intercept any of the underground channels.

Since one cannot see beneath the surface of the ground, the drilling of wells was always an uncertain operation, and in their doubt about where to dig for water farmers often consulted "water witches" or "dowsers". Such people were supposedly endowed with supernatural powers that enabled them to discover the location of underground streams. A dowser generally walks about with a forked stick, tightly held, which is supposed to dip violently when he crosses the position of an underground stream. His success, if any, has no known scientific basis.

Perched Water

An aquifer may rest on an impermeable substratum that

overlies unsaturated material above the normal water table. The water in such an aquifer is *perched*. It is prevented from percolating downward to the normal water table by the impermeable material beneath. Bodies of perched water are especially common in arid or semi-arid regions.

Confined Water; Artesian Wells

A permeable sediment, such as an uncemented sandstone, may be overlain by an impermeable one, such as shale. If the rocks have been tilted and eroded, the permeable sediment may be exposed at the surface in hills or mountains that rise above the level of the overlying impermeable shale nearby. Under these conditions the ground water in the aquifer may be both *partially confined* (in the area where it is overlain by shale) and *easily replenished* (in the area where it appears at the surface). Water enters the aquifer at A where it is exposed at the surface—the intake area.

The water table (t-t2) within the aquifer of the intake area is higher than the ground surface at B, where the aquifer is confined beneath overlying shale. If a well (W) is put down to this aquifer in area B, the confined water in the aquifer will rise through the well under the pressure of the head of water from the intake area and flow out on the surface of the ground, making this an *artesian well*.

The name "artesian" was originally restricted to flowing wells. It is now applied, however, to any well in which the water rises above the elevation of the aquifer penetrated.

GROUND-WATER MOVEMENT

The water that fills the pores below the water table is not static. It moves slowly under the influence of gravity. The direction of movement is controlled by the form and slopes of the water table. From the higher points on the table the water moves laterally, vertically, or obliquely downward. Similarly, it flows from all directions toward troughs or low points. The overall tendency is to lower the high points on the water table and either to raise the low points or else discharge water from them.

Points of surface discharge, located where the water table intersects the ground, are called *springs*. Most perennial streams also mark areas of discharge in the water table. They lie in troughs on the water table, and the ground-water flow is toward and into them. Such streams are called *effluent* with respect to ground water. Streams that flow from mountains or other well-watered areas into deserts or semi-arid regions, however, may lose water to the water table. Such *influent streams* lie on ridges in the water table.

If permeability and other factors are uniform, movement between two points on the water table is determined by the *hydraulic gradient*, which is the ratio between the difference of elevation, or head (H), and the horizontal distance between the two points (L). Ground-water gradients are usually low, such as a fall of 1 foot per 1,000 feet (0.001) or 10 feet per 1,000 (0.01).

Effect of Variations of Intake on Water Levels

For any particular permeability, the hydraulic gradient adjusts itself to the supply of water. If the discharge, as into streams, is temporarily greater than the supply into the intake area, the water table flattens. Thus, in dry spells the water table sinks farther below the surface under the groundwater divides, reducing the hydraulic gradient and the discharge.

In semi-arid western Texas, eastern New Mexico, and adjacent parts of old Mexico the water table is relatively flat and, in many places, lies 500 to 1,000 feet below most of the land surface. Much of the region either lacks permanent surface streams or else the streams are on perched ground-water bodies.

DARCY'S LAW

The modern concepts of ground-water movement were discovered early in the Nineteenth Century. Almost all movements of ground water are so slow that they occur by laminar flow, whereas practically all movements in surface streams are turbulent. Flow lines in laminar flow are smooth, continuous, and traceable. Investigation of the flow of water in pipes shows that for turbulent flow the velocity, and discharge are approximately proportional to the square root of the hydraulic gradient, whereas *in laminar flow the velocity and discharge vary directly as the hydraulic gradient.* This fundamental law, as it applies to groundwater movement, was discovered and formulated in the 1850's by the French hydrologist Henry Darcy, during his study of the water supply of the city of Dijon. Darcy's law may be stated as:

$$V = PI$$

where: V = velocity of ground water P = coefficient of permeability—a constant determined by the character of the material through which the water moves—a measure of the ease with which water moves through a subsurface material I = hydraulic gradient—the slope of the water table

Geologists are usually more interested in the quantity of water moving than in its velocity. Hence, we usually replace the velocity term in Darcy's law by an equivalent term which states the quantity of water moving per unit of time (Q) through a given cross-sectional area (A).

and we write Darcy's law as or

$$Q = PIA$$

where: Q = quantity of water moving per unit of time (measured in gallons per day, cubic meters per day, etc.) A = cross-sectional area through which water moves

If A is expressed in square feet, if I is made 1 (100 per cent gradient), and if Q is measured in gallons per day, then P comes out in *Meinzer units*. Those aquifers that have been investigated in this way have permeabilities that range all the way from 10 to 90,000 Meinzer units. The approximate yield of a well may be predicted if the permeability of the aquifer is known, together with the hydraulic gradient under pumping conditions, and the cross-sectional area (the area of perforated casing through which water is entering the well). Similarly, the amount of ground water passing through a cross section of an aquifer can be calculated.

RATES OF GROUND-WATER MOVEMENT

The movement of unconfined ground water through uniformly permeable material, a section showing both ground-water divides and effluent stream levels. Some flow lines go deep, but little water follows these. The maximum velocities (and hence the greatest volumes of water transferred) are near the streams where the hydraulic gradients are steepest.

The actual velocities of ground-water movement, though almost everywhere low as compared to stream velocities, are highly variable. They vary in different parts of a single body of water moving through rock of uniform permeability.

Mean or average rates of movement can be calculated most easily from Darcy's Law, if the coefficient of permeability is known. The late leader of American hydrology, O. E. Meinzer, considered the flow of 50 feet per year in the Carrizo sandstone of Texas to be typical of many aquifers. In places much more rapid movements have been demonstrated by use of dyes or salts, introduced at one observation well and recovered or measured indirectly at another. Movements of 10 or 20 feet per day are sometimes attained in highly permeable materials, and one extreme velocity of 420 feet per day has been reported.

Balance Between Discharge, Permeability, and Hydraulic Gradient

According to Darcy's equation, in materials of low permeability the gradient of the water table increases steeply as intake water is added locally to the mass of ground water. In materials of high permeability the water table is relatively flat, and hydraulic gradients may be only a few feet or a few tens of feet per mile.

Discharge of Ground Water into the Ocean

The discharge of fresh ground water directly into the ocean may produce a body of fresh water extending far below sea level beneath an oceanic island or along a continental margin. It appears as if the lighter, higher column of fresh water were in static balance with the heavier sea water like a foreign mass floating within it. A column of sea water 1,000 feet high can balance one of fresh water about 1,025 feet high. Thus a column of fresh water rising to an elevation of 25 feet above sea level, near the ocean shore, might indicate that fresh ground water could be recovered to a depth of 1,000 feet below sea level. Such a water body is, however, not static; it is constantly discharging into the sea. The friction of flow through the pores retards the spreading out of the fresh water, but if it were not constantly replenished from rainfall, the vertical separation between salt and fresh water would soon disappear. The probable relationships and flow lines in such a water body may be deduced from Darcy's Law.

The body of fresh water under Oahu, one of the Hawaiian Islands, though hundreds of feet thick, is thinner than anticipated on the static flotation hypothesis. Furthermore, the fresh water is underlain by a thick transition zone of brackish water. Perhaps the mixing is due in part to oscillations caused by the intermittent pumping from many large wells. The original accumulation of fresh water extending far below sea level is presumably the result of flow following the lines induced by the hydraulic gradient of the fresh-water surface.

Drawdown by Pumping

A pumped well producing from an unconfined aquifer disturbs the water table by setting up a point of artificial discharge. We have already seen how the water table becomes adjusted to points and lines of natural discharge, such as springs and effluent streams. Similarly, removal of water through a well draws the water down adjacent to the well to produce a *cone of depression* in the water table that greatly increases the hydraulic gradient close to the well.

The pumping well removed water from uniformly permeable alluvium in the valley of the Platte River. The undisturbed water table sloped eastward 6 or 7 feet per mile. The lowering of the water levels was determined in more than 80 observation wells drilled in 8 radial lines extending out to about 1,200 feet from the pumping well is a section across the "cone" of depression in the direction of slope of the undisturbed water table. Note that a lowering, or *drawdown,* of 32 feet at the pumping well after 48 hours of pumping was accompanied by a lowering of 1 foot in the water table at a distance of 250 feet, and that measurable lowering extended at least 1,200 feet from the pumping well. Flow lines toward the well must have extended equally far laterally, and also far below the water table. Effects analogous to the cone of depression may also be observed in wells that tap a confined aquifer. If a group of artesian wells is capped so that no water escapes, the water in each well will exert a definite pressure against the top of the casing. If the cap of one well is opened so that the water begins to flow from it, a decline in pressure will soon be noted at all the other wells.

COMPOSITION OF GROUND WATER

Rain water, if it passes slowly through limestone, or even through decaying moderately calcium-rich rocks, such as granite, may pick up enough calcium ion—($C_a CO_3 + H^+$ '! $Ca^{++} + HC0_3$-)—to become *hard water*. The amount of calcium ion in hard waters in humid regions is only a small fraction of one per cent. Such waters may contain as much sodium ion as calcium ion, but the sodium ion is not easily precipitated and ordinarily goes unnoticed.

In arid regions, however, water within a few feet of the earth's surface may be partially or completely evaporated after each rain, thus progressively concentrating the water-soluble salts in the soil water. The commonest salts thus concentrated are the carbonate, chloride, and sulfate of sodium. The ground water in many arid regions is, therefore, more or less *alkaline*. Alkaline water is very toxic to plants, and the "alkali soils" are nearly useless agriculturally. If such areas are drained, and the alkaline water flushed downward by heavy irrigation, they may be reclaimed for agricultural use.

Still other ground waters are *salty*. Such waters contain enough sodium chloride to make their taste unpleasant and to make them injurious to plants. Some salty ground waters are in whole or part sea water, infiltrated directly from the ocean. Others have dissolved salt from salt beds in the rocks. Some water in deeply buried marine sedimentary rocks is supposedly sea water entrapped at the time of their deposition. Such entrapped sea water is called *connate water*. Connate waters rarely have exactly the composition of the present ocean. Variations are probably mostly due to dilution since burial, to concentration by evaporation previous to burial, and to chemical reactions with the enclosing rocks.

GROUND WATER IN CARBONATE ROCKS

Solution Passages

Rain water is so effective a solvent of carbonate rocks that in limestone regions it enlarges cracks and pores and dissolves tunnels, irregular passages, and even large caverns along joints or other openings in the rock. Solution of the rock may become so extensive that much of the surface drainage goes underground through vertical tubular passages called *sinks* and discharges through *caves.*

The development of sinks and caves is, of course, a very slow process. Water percolating down a crack in limestone enlarges the opening by solution. Weathering, rainwash, and collapse of the walls by gravity widens the opening at the surface, thus enabling it to trap more and more surface water which, in turn, causes more rapid solution of the limestone walls. The vertical sink thus formed may extend completely through the limestone bed, discharging at its base into other caves and caverns formed by solution of the limestone. In time, the entire bed of limestone may thus become honeycombed with interconnected sinks and caverns. Within the caves a part of the dissolved calcium carbonate may be redeposited, often in bizarre iciclelike forms, producing masses of dripstone.

The easy movement of water through a cavernous limestone aquifer and the flatness of its water table have been often demonstrated by pumping tests in mines. At the Los Lamentos mine, in Mexico, about 100 miles southeast of El Paso, Texas, two years of constant pumping did not lower the water table appreciably, and the mining of a rich ore body was necessarily stopped at the water table.

Karst Topography

In some limestone or dolomite regions there are few or

no surface streams, and the water table is flat and far below the surface. The rainfall passes underground at once through sinks and joints widened by solution, and flows in underground streams with low gradients through large and small caverns, cascading at intervals to still lower levels, until, finally, it reappears at the surface as giant springs on the walls or floors of deep valleys. Such an area of sinks and underground drainage is called a *karst* region, from the name of such an area in Yugoslavia. Other well-known karst regions are the Causses of southern France west of the Rhone River and parts of the Cumberland Plateau in Kentucky and Tennessee.

The topography of a karst region differs from that of an area with normal surface drainage. Instead of a system of slopes definitely integrated to surface streams, a karst region is pock-marked by a maze of large and small depressions. The smaller depressions are the upper ends of sinks and other solution passages, enlarged at the surface by weathering, rainwash, and downslope movements. Larger openings appear where the roofs of caverns have collapsed. Through such an area the main rivers flow at the bottoms of deep valleys, gaining most of their increase in volume from numerous large springs rather than from tributary streams. The largest springs arise from solution channels or even extensive caverns, which, in turn, are fed by surface drainage that goes underground. A stream may enter a sink in one valley and emerge through springs in a neighboring valley. As a result, many valleys are dry for short or long stretches.

Karst drainage has influenced men's actions for ages. Life is hard on the plateaus, easy in the well-watered valleys. Near the Causses of southern France great springs that emerge at the edge of the Rhone Valley have determined the sites of towns. Many of these, such as the ancient Roman city of Nîimes, have flourished since remote antiquity.

Ground Water in Volcanic Rocks

The cracks and cavities characteristic of many masses of volcanic rocks, as in the Hawaiian Islands or the Columbia

Plateau of the Pacific Northwest, make these rocks almost as permeable as cavernous limestone or dolomite. The most productive group of springs in the United States is in volcanic rocks. The springs are distributed along a 50-mile stretch of the Snake River, Idaho, and discharge 14 million cubic meters of water per day, or 5,000 cubic feet per second, contributing to the Snake a volume of water two-thirds as great as the average discharge of the Mississippi River at St. Paul, Minnesota.

SOLUTION AND CEMENTATION BY GROUND WATER

In general, solution is more important above the water table; deposition and cementation below. The extreme effects of solution in carbonate rocks have already been described. Solution is not confined to carbonate rocks; in the long lapse of geologic time even minerals that the chemist regards as highly insoluble are appreciably dissolved. Grains of garnet in sandstones may be pitted and etched by ground water, and such highly insoluble minerals as augite and hornblende may be completely dissolved. Fossil shells composed of calcite are commonly leached out of shales and sandstones, leaving open cavities that may be filled to form "casts" which preserve faithfully the fossil form. Casts of soluble minerals such as halite or pyrite also are common.

The cementation of sand to form sandstone was mentioned. Calcite cements many sandstones but the reason it deposits in the sandstone pores is obscure. Evaporation of water and resultant deposition of calcite-the mechanism of dripstone formations—can hardly operate below the water table. Possibly one factor is the release of pressure on rising bicarbonaterich ground water allowing the escape of carbon dioxide and the resulting precipitation of calcite. Around shell fragments, some sandstones are locally cemented, forming ball-like masses called *concretions*. This suggests that the shell fragments were nuclei upon which calcite from the saturated ground water precipitated.

Ground water may also deposit other substances. Silica,

as opal or quartz, and iron oxide, as limonite or hematite, may fill or coat cavities in rocks. At depths of two or three miles temperatures are close to the boiling point of water at the earth's surface, and many other minerals may form in the pores.

Ground-Water Supplies in the United States

Water supply is a primary concern of farmers and inhabitants of arid regions. Even in the humid but populous areas of western Europe and eastern North America industrial demands put a heavy strain on water supply. The total quantity of recoverable fresh water underground at any one time is a matter of interest, especially in regions dependent on ground water for their supplies. All estimates indicate that the total amount of ground water is vastly less than the amount of water in the oceans, but much more than that in the atmosphere, or than that which falls as rain and snow in a single year. In almost any area some water can be obtained from wells, but in nonporous rocks, even where shattered, yields are relatively small, and of course yields will be negligible from nonpermeable rocks even if they are porous.

Abundant yields of ground water come chiefly from unconsolidated surface formations, mostly of Pleistocene and Recent age, and secondarily from older consolidated, but still permeable, sediments or from lavas. The principal unconsolidated aquifers are:

- Alluvial gravels and sands that fill deep interior basins,
- Glacial outwash sands and gravels,
- The coarserparts of deltaic and other coastal plain deposits, and
- Sands and silts beneath river floodplains.

Interior basins filled with unconsolidated sediments are common in the western third of the United States. Ground water from many of them has been extensively developed because of the general aridity of the region. They normally supply about one half the ground water used in the United States. Several large basins are in California, including the large central valley (basins of the Sacramento and San Joaquin rivers), which is 450 miles long and 35 miles wide; and the much smaller valley of southern California, which includes the coastal plain near Los Angeles. In 1948 the ground water extracted from these basins in California was enough to cover 9 to 10 million acres to a depth of 1 foot—over 35 per cent of all the ground water used in the United States.

Glacial outwash sands and gravels are interbedded with, or lie slightly south of, the relatively less permeable Pleistocene till sheets. Glacial gravels grade into river floodplain and deltaic gravels and sands, notably along the Mississippi River. Large floodplain aquifers are also found on the Great Plains and permeable coastal plain deposits extend from New Jersey to Texas.

Among older, more consolidated rocks the chief aquifers are:

- Permeable sandstones,
- Well-jointed volcanic rocks,
- Cavernous limestone or dolomite, and, more rarely,
- Fissured crystalline rocks such as quartzite or shattered granite. Sandstone aquifers supply large amounts of ground water in the Dakotas and elsewhere in the northern Mississippi Valley but are important also in Texas, Michigan, and other regions. Basalt flows are important aquifers in the Pacific Northwest and in Hawaii. Cavernous limestones yield abundant ground water in Florida and in parts of the Cumberland Plateau.

The aquifers of three highly productive areas will be described in the following pages. These examples include the glacial outwash materials and Cretaceous sands of Long Island, the sands and gravels in the valley of southern California, and the Dakota sandstone of the northwestern Mississippi Valley.

GROUND WATER OF LONG ISLAND

Long Island has no large streams, partly because of its size and partly because infiltration is so easy. The 40 to 50 inches of annual precipitation in west central Long Island, it has been estimated, is taken to surface runoff, ground water, and transpiration 20, 40, and 40 per cent, respectively.

The most productive Cretaceous sand is the basal bed, 100-250 feet thick. It is a clean quartz sand, practically uncemented and confined by an overlying shale. The Cretaceous sands and shales are unconformably overlain by Pleistocene sediments as much as 300-400 feet thick. Interbedded Pleistocene till, sand, and gravel form two ridges, one near the northwest shore and the other along the middle of the island. The outwash plains of sand and gravel lie between and south of the ridges.

In a small Brooklyn area excessive pumping lowered the water table until it was below sea level. The water table was 0-15 feet above sea level (0-50 feet below the ground surface) in 1903, but had been drawn down to a maximum depth of 34 feet below sea level by 1943. This reversed the hydraulic gradient, and salt water from the sea invaded the margins of the aquifer. The State of New York thereafter has required water pumped from new cooling and air-conditioning wells to be returned to the aquifer from which it was withdrawn. By 1946 more than 200 recharge wells were returning water underground in the urban portion of Long Island. In the rural areas several large surface recharge basins, where storm and industrial waste water could be stored until it would seep into the ground, had also been established. Total recharge in the summer of 1944 amounted to 60,000,000 gallons a day. The warmer recharge water raises the ground-water temperature a very few degrees Fahrenheit, making the water less valuable for cooling. The recharge wells and storm-water basins have served their main purpose, however; the water table is no longer falling, and salt-water inflow has decreased.

Ground-Water Basin

In semi-arid southwestern California, several large interior basins have been filled with alluvium, across which the intermittent surface streams flow to the sea, charging or recharging the alluvium with ground water on the way. The basin is filled with moderately well-sorted alluvium, 500 to 1,200 feet thick. Most of the alluvium came from the northeast and has been deposited in alluvial fans that have united to form a compound apron extending almost entirely across the whole basin. Coarse gravels are present at the heads of the fans, where the slope of the surface and initial dips of the strata are as high as 9°. Farther out on the fans the gravels grade into, and are interbedded with, sands and silts. Near the fan heads relatively impermeable soils have formed, several of which have been buried by upbuilding of the fans. The permeable gravels and coarse sands make a rather complex set of aquifers. On the lower portions of the apron some wells were originally artesian, the aquifers being confined by finer sediments or soil zones, and the hydraulic head being the result of the initial dip.

Many ground-water basins in California are broken by faults so recent that they cut through the alluvium. In this area the faults are effective barriers to ground-water movement because impervious clay gouge has been smeared along the fault planes. The water table drops 400 feet—a vital difference to those who must obtain their water from wells.

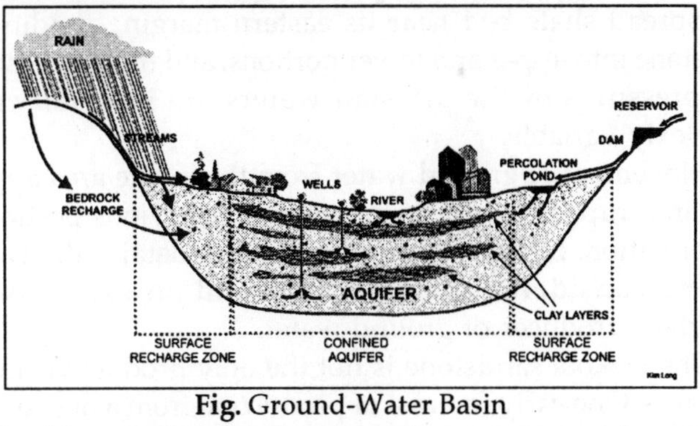

Fig. Ground-Water Basin

In this populous but arid region an attempt is made to keep the winter flood waters from reaching the sea by diverting them onto complex spreading grounds of coarse gravel which provide opportunity for the water to infiltrate. These spreading grounds serve the same purpose as the recharge wells and basins on Long Island. It should be noted that lowering water levels in a basin by pumping is not wholly a misfortune. Surface or subsurface outflow from the basin is decreased or even stopped, and the subsurface reservoir is partially emptied, and so prepared to take in a much larger recharge.

DAKOTA SANDSTONE ARTESIAN AQUIFER

The great Dakota sandstone artesian basin is the largest and most important source of artesian water in the United States. It extends over much of North Dakota, South Dakota, Nebraska, and parts of adjacent states. At least 15,000 wells have been drilled into this Cretaceous sandstone, which is generally somewhat less than 100 feet thick, and is overlain by hundreds or even thousands of feet of other sediments, mostly impermeable shales.

The basin form resulted from folding. The principal intake areas are at the west, in the upturned zone along the edges of the Black Hills and Rocky Mountains, 2,000 or more feet higher than the sandstone on the east side of the basin. In general, there is a hydraulic gradient from west to east. Pressures have decreased progressively since the first well was drilled in 1882. The sandstone is, however, not a simple aquifer. It includes a widespread shale bed near its eastern margin, dividing the sandstone into upper and lower portions, and the compositions and pressures of the artesian waters are in detail rather complexly variable.

Movement of ground water from the intake area appears to be interrupted locally by relatively impervious portions of the formation. Despite these complicating details, the Dakota may be considered a single unit, and an exceptionally important producer of ground water.

The Dakota sandstone is not the only productive aquifer in its area. One early forecast of production from a deeper sand is still remembered. The early railroads crossing this and region were large users of water.

Near the Black Hills the Dakota sandstone was an inadequate source. In 1905 N. H. Darton of the U. S. Geological Survey, who had just completed an investigation of ground water in the Great Plains region, recommended that the

Burlington Railroad go deeper for water in the area near the Black Hills.

The structure sections and geologic maps he had made indicated that by drilling at the town of Edgemont a probably productive Paleozoic aquifer would be encountered at a depth of about 3,000 feet. After almost three years of old-fashioned impact drilling a well flowing more than 400,000 gallons a day was developed at a depth within 31 feet of that predicted. Forty years later a new well, completed in the same aquifer after 50 days of rotary drilling, had an initial flow of almost 1½ million gallons per day, an unusually large yield from a Paleozoic sand, and very precious in this dry country.

Economic and Legal Aspects of Ground-Water Use

Where there is not enough surface and ground water to go around, disputes have arisen between individuals, communities, and states, and these disputes have been carried to the courts. Communizing the rule that the owner of the land surface also owns everything beneath the surface, all the owners of land above a ground-water basin are considered to have joint ownership of the basin.

Water may not be exported without good reason beyond the basin's surface area. In addition, the *principle of best use* has been established, as in a dispute between cattlemen, who wish to preserve feeble springs, and truck gardeners, who wish to put the abundant ground water of the same area to more productive use.

In some states, such as New York and Maryland, the permission of state authorities is required for the drilling of large wells, and return of used waters to the aquifer may be required. Ground water is an important public commodity, and its use now requires regulation by well-informed public officials.

Chapter 4

The Oceans

OCEANOGRAPHY

Fossil Corals weathered from outcrops of limestone in the desert and marine shells in the rocks at the summit of high mountain ranges prove that the relative positions of land and sea have changed in the geologic past. If we are to read the history of the earth from the record of the marine rocks, we must know something of the sea, the processes that go on within it, its organisms, and its varying environments of erosion and sedimentation.

The study of the sea—the science of *oceanography*—is young. Its modern development began with the great expedition of the British research ship "Challenger" in 1873-1876. Rapid advances in our understanding have been made, but the vastness of the oceans and the complexity of their problems still leave many puzzles unsolved—puzzles whose solution would greatly aid our interpretations of the geologic record.

Composition of Sea Water

Except where great rivers pour into the sea, the chemical composition of sea water is almost uniform. The concentration of the dissolved ions varies slightly from place to place because of differences in evaporation and precipitation, but their relative proportions are everywhere nearly constant. Mixing and diffusion, although sluggish and by no means complete, have been very effective during the long history of the oceans.

Average sea water contains about 3.5 per cent of dissolved

matter—35 parts per 1,000. More than 40 elements have been detected in sea water. A few of these overwhelmingly predominate, but several minor constituents, such as phosphorus and nitrogen, deserve special mention because they are important to life.

It is assumed that the dissolved solids were brought to the sea by rivers. The runoff from the lands has been carefully estimated, and many analyses of river waters have been made. From these data, the American chemist F. W. Clarke has computed that about 3 billion tons of dissolved solids are brought yearly to the sea. These solids, however, are very different from those composing the oceanic salts.

Table: Composition o Dissolved Solids in River Water and in the Sea

Ion	*Rever Water (Weighted Average)*	*Sea Water*			
CO_3^{-}	35.15%	0.41 (HCO3—)%]			
SO_4^{-}	12.14	7.68			
Cl^{-}	5.68	55.04			
NO_3^{-}	0.90	—			
Ca^{++}	20.39	1.15	Mg^{++}	3.41	3.69
Na^{+}	5.79	30.62			
K^{+}	2.12	1.10			
$(Fe, Al)_2 O_3$	2.75	—			
SiO_2	11.67	—			
Sr^{++}, $H_3 BO_3$, Br^{-}	—	0.31			
	100.00	100.00			

Either the river waters of today differ greatly from those of the past, or else oceanic processes are selectively removing certain constituents from the sea. River waters differ from oceanic waters chiefly in their great excess of carbonate, calcium and silica and lower content of sodium and chlorine. The calcium contributed by the rivers would double the concentration in the sea in about 100,000 years, if it were not precipitated. Obviously, silica must also be quickly precipitated. These substances are extracted by marine organisms to form extensive deposits of rock.

Circulation of the Sea

Water, like all fluids, has no strength; it flows in response to the slightest differences in unbalanced pressure. If we heat a pan of water, the warmed water expands—that is, it becomes less dense—and rises, whereas the cooler, heavier water sinks and flows beneath it. Similarly, if we place pure water in one compartment of a vessel with a removable partition, and salt water at the same temperature in the other compartment, on removing the partition the heavier salt.water will flow beneath the fresh. For a time the water will be layered according to density, but diffusion ultimately makes it uniform, and the layering disappears. Differences in temperature and salinity cause differences in density in the ocean and set up a circulation of the ocean water.

Though less effective than that of the air, the circulation of the sea notably affects climates. Changes in the configuration of land and sea, such as are implied by the distribution of marine rocks, must have affected ancient climates also, with all that this implies in respect to erosion, deposition, and life environments.

DENSITY VARIATIONS IN THE SEA

Oceanographers have found that the major circulation of the sea is a response to two main factors: (1) the drag of the prevailing winds and (2) the differences in density of different water masses. Both factors largely depend on the sun's radiation. Density differences are generally controlled by temperature and salinity (concentration of dissolved salts), but the local content of micro-organisms or of suspended sediment also modifies these effects.

The oceans are so vast that diffusion only slowly equalizes differences in salinity brought about by evaporation and precipitation. Similarly, the heat conductivity is so low that temperature differences persist. Hence the oceans are made up of great water masses within each of which the temperature and salinity vary but little, though the boundaries against neighboring masses with differing characteristics may be relatively sharp. The differing densities of these water masses and the systematic drag of wind friction on the surface of the sea sets up a slow motion and causes the oceanic circulation.

Variations of Density with Latitude

As there mentioned, most is carried by the winds, but in both the North Atlantic and North Pacific oceanic currents carry about 10 per cent of the total heat transported.

Water strongly absorbs solar radiation; even the clearest ocean water absorbs more than 80 per cent within 33 feet of the surface (more than 60 per cent in the first 3 feet). Water turbid with silt, air-bubbles, or micro-organisms absorbs even more—over 99 per cent within the first 33 feet. The sun thus warms only a thin surface layer of the oceans.

Warming the surface water layer makes it less dense, but warming also increases evaporation and thus tends to increase its salinity and, hence, its density. In the equatorial belt evaporation is high, but wind speeds are low, cloud cover is dense, and rain abundant. Rains dilute the surface water, counteracting evaporation in its effect on density. Farther from the equator in the trade wind belts, despite lower temperatures, evaporation is higher because here are strong winds, few clouds, and little rain. Here, a surface layer about 2,000 feet thick is so warm that it floats on colder water despite its high salinity.

The interplay of evaporation and precipitation on salinities. Because of differences in temperature and salinity the warm water of the equatorial and trade wind belts tends to spread north and south over the colder water near the poles. Conversely, the cold polar water sinks and flows equatorward. These movements have been traced by measuring salinities and temperatures at many places in the ocean. In the Pacific Ocean water that sinks in the Antarctic can be traced by its distinctive temperature and salinity even to the equator; cold Arctic water can also be traced far to the south in the depths of the Atlantic. The slowly drifting water masses are so large that mixing fails to disperse them completely.

Warm surface waters, however, do not by any means flow directly from the equator toward the poles. It is clear that density differences alone do not explain their circuitous paths. Deflection of the masses by wind friction is one obvious factor; another is the rotation of the earth.

Effect of the Earth's Rotation

The tendency of a moving mass to persist in its line of motion (its inertia) causes the trade winds to be deflected to the right (westward) in the northern hemisphere and to the left (also westward) in the southern. The same relations prevail with ocean currents, but because the sea, unlike the air, is confined to depressions on the earth's surface, and also because its density and hence its inertia are higher compared to the driving force, its movements show a somewhat more stable pattern. There are seasonal displacements of currents, but these are much less capricious than those of air masses. In general, the density differences in the sea arise in the same places, and the motion that tends to equalize them becomes a steady flow.

Pressure Differences in the Sea

The fundamental cause of movement of the large water masses in the sea is the differences in density set up by variations in temperature and salinity. Such differences cause unbalanced pressures that operate in the vertical direction, at right angles to the earth's surface. On the other hand, the force set up by the rotation of the earth operates in the horizontal direction and it changes in amount with differences in latitude. Wind friction also sets up horizontal forces. From detailed observations of temperature and salinity, made at measured depths and at points where longitude and latitude are known, it is possible to calculate the resultant pressure existing at any definite level beneath the surface from all of these forces. *Isobaric maps* (isobars are lines of equal pressure) can thus be prepared for the sea, just as they are from barometric readings of the air.

Since water has no strength, it is also possible to calculate the currents that will be set up and maintained by the pressure differences. The calculations involve hydrodynamical principles beyond the scope of this book, but one of the triumphs of modern oceanography has been the close agreement between the current motion as determined by measurements and the motion as calculated from the theory of flow. The method has, therefore, become standard practice in oceanography, because it is much easier to measure temperature and salinity than the velocity and area of the moving water masses.

One general rule emerges from the analysis of pressure differences: In the Northern Hemisphere, if the observer looks in the direction that the current is traveling, he sees lighter water on his right and denser water on his left—the isobaric surface slopes upward to the right. In the Southern Hemisphere, the lighter water lies to the left, and the denser water to the right. As the density of the water near the ocean surface is more dependent on temperature than on salinity, "lighter water" may, in general, be replaced by "warm water," in the rule stated above. Let us turn to a description of one of the largest and best known of the oceanic currents, the Gulf Stream.

THE GULF STREAM SYSTEM

"There is a river in the sea—the Gulf Stream," said Maury, the great American oceanographer of a century ago. This statement was regarded by some as a typical American exaggeration, but modern studies essentially confirm his picture of a well-defined stream of warm water, with relatively sharp boundaries, coursing with the speed of a river across thousands of miles of ocean.

The Gulf Stream system consists of three segments: the Florida Current, the Gulf Stream proper, and the North Atlantic Current. The Florida Current pours through the Florida Strait with a speed of nearly 3 miles per hour, carrying as much water out of the Gulf as 14 Mississippi Rivers would bring into it. Theoretically, sea level should stand about 18 inches higher on the Cuban coast than at Key West, Florida. Of course, this cannot be measured, but measured current speeds in different parts of the channel agree with the speeds computed from the measured temperature and salinity.

Temperature and salinity determine the water density from which velocity is computed by physical theory.

Precise leveling across Florida shows that the Gulf of Mexico stands about 7 inches higher than the Atlantic—this is the immediate cause of the Florida Current. The higher level in the Gulf is, in turn, caused by the westward deflection by the trade winds of the warm surface waters of the Atlantic. Most of the water feeding the Florida Current flows northward into the Gulf through the Yucatan Passage between Yucatan and Cuba.

As the Florida Current emerges from the Gulf of Mexico it is joined by a current which flows north along the east side of Cuba. As the combined stream sweeps north along the continental shelf it is continually augmented by water from great eddies in the Atlantic.

Deflected to the right by the earth's rotation, the stream leaves the continental shelf and courses northward as a well-defined broad current. Its speed is as great as 2½ miles per hour, and its depth, hitherto about 2,600 feet, becomes as much as 6,500 feet. As it sweeps far offshore past Virginia, it is about 60 miles wide and in volume about three times that of the Florida Current—in other words, that of 40 Mississippis. Inshore from the Gulf Stream, great counterclockwise eddies are thrown off, giving rise to a southward drift along the Atlantic shores that shifts sand southward along the beaches, and may help to shape the cuspate forms of the sandy capes from Hatteras southward.

The Gulf Stream spreads out to the northeast and becomes less definite. It gives off great eddies both to the north and south and encloses still smaller ones. Here, it is called the North Atlantic Current. Its northern branches spread far into the Norwegian Sea and have been traced beyond North Cape into the Arctic Ocean. They notably affect the climate of the British Isles and Scandinavia, for they transport warm water far to the north.

Effects of the Shape of the Coast on Currents

Coastal configuration strikingly affects ocean currents,

especially along the eastern shores of the Atlantic and Pacific. The Benguela Current along the west coast of Africa, the Peruvian Current, and the California Current are all in part maintained by the prevailing winds which blow equatorward, either parallel to the shore or offshore. The wind drift of surface waters is, therefore, offshore along all three coasts, and, as no surface flow is available to replace the water lost, each of these currents is replenished by upwelling of colder water from the depths. Such deep water is rich in plant nutrients and supports an abundant marine life.

With climatic fluctuations the currents are occasionally disturbed, as has happened several times with the Peru Current. The upwelling may then cease for a time. As a result, the marine life may be destroyed on a wholesale scale, and a rain of dead organisms falls to the sea floor. Perhaps some fossil-rich strata record similar disturbances of oceanic currents in the geologic past. Such disturbances of currents may, in turn, give rise to local climatic changes.

When the Peru Current shifted, torrential rains wrought tragic destruction to the normally arid west coast of Peru and Ecuador. Changes in configuration of sea and land during the geologic past must also have affected the ocean currents. For example, if the Isthmus of Panama were submerged (as we know from marine sediments that it has been), or even if the eastern point of Brazil were several degrees north of its present position, the power of the Gulf Stream would be greatly changed. Water warm enough for the growth of coral reefs now bathes the Bermuda shores; it may be doubted that a feebler Gulf Stream would allow these animals to grow so far north. It must be kept in mind, then, that the environment of modern marine organisms depends not only on climates determined by latitude, but also on the courses of ocean currents.

Shore Processes

The great ocean currents intimately affect climates and marine life, but the oceanic agencies that most directly affect the shore are the waves, tides, and longshore currents associated with them. Waves and wave currents wash every shore on earth. Tides, though locally spectacular, are generally far less effective in shaping the shores, though incomplete information suggests that they are important in controlling the distribution of marine sediments at considerable depths.

WAVES

Geologic Effectiveness

Anyone who has systematically watched the never-ending play of waves on the shore—whether the gentle ripples that carry grains of sand up and down the beach or the great rollers that pound against the cliffs—cannot but be impressed with the power of the sea in molding the shape of the coast and in moving detritus. Historic records prove that for decades, or even for centuries, some shorelines have been driven back several feet per year by wave attack. From repeated soundings and studies of bottom sediments we also find that waves profoundly affect the movement of sediments at shallow depths.

The Theory of Waves

The restless waves of the sea seem almost to defy analysis, and, indeed, no theories yet proposed explain all their vagaries. Nevertheless, the need—vital alike to the naval architect designing ships, the civil engineer building docks and jetties, the military scientist planning amphibious landings, and the geologist studying sedimentation and erosion —has led to decades of intensive study and many useful conclusions.

The mechanism by which wind forms water waves is complex and not completely understood. Theoretically, waves should not arise until the wind speed is 11 to 15 miles per hour, yet measured winds of only about 2½ miles per hour produce waves on ponds. Clearly, wind motion must be turbulent, that is, some air particles move at much higher speeds than the current as a whole. These variations in wind speed produce corresponding variations in surface pressures and friction; it must be these that first ruffle the surface.

Now, it has long been known from watching floats that the water particles in surface waves move up and down in orbits, but retain their average positions. Indeed, if the water masses actually moved forward at the same speed as the wave, no ocean would be navigable.

Wave Characteristics

A wave is described in terms of its length, L (the distance from crest to crest or from trough to trough); its height, H (the vertical distance from trough to crest); and its period, T (the time interval between the appearance of two consecutive crests at a particular position). The velocity, V, of a moving wave is the distance the wave form travels in a given time; thus, $V = L/T$.

The motion of the water particles, and hence the wave form, depends on the wave length, the ratio of height to length, and the depth, d, of the water. Waves in water that is deep compared to wave length (d is greater than $L/4$) are called *deep-water waves*. Their speed is independent of the depth.

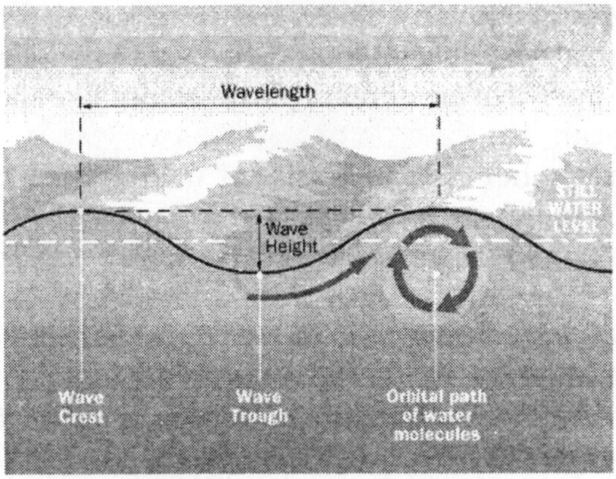

Fig. Schematic Diagram of a Progressive Wave.

The water particles in a deep-water wave move approximately in circles, the radii of which decrease very quickly with depth. For waves 330 feet long and 16 feet high, traveling about 28 miles per hour, the particle velocity at the surface is about 4.4 miles per hour; at a depth of 65 feet about 1.2 miles per hour; while at 330 feet the particle motion is negligible. This quick falling off of orbital velocity with depth gave Vening-Meinesz' submarine enough stability for gravity observations in moderate seas when it was submerged less than 100 feet.

But in great swells 1,250 feet long and 33 feet high, traveling at 56 miles per hour, though the particle velocity at the surface is again only about 4.4 miles per hour, the motion persists to greater depth and at 330 feet is still nearly 1 mile per hour.

The water particles do not move at exactly uniform speed in their orbits; they are slightly faster when at the top of the path. Here they are moving in the direction of wave progress, so there is a slight net forward movement of water with the wave. For low waves of long period this net movement is slight but it is noticeable in high steep waves.

Under steady winds the waves grow in size and speed, up to a limit. The wind gives them energy in two ways: by its push against the wave crests and by the frictional drag of the air along the water surface. The first process depends on the difference between wind speed and wave speed; when wind speed is greater the waves are pushed, but when wave speed is greater the wave encounters air resistance like that against a moving car.

On the other hand, frictional drag depends not on differences in the speed of air and the wave *form*, but on the differences of speed between air and water *particles*. As we noted above, the particle speed is moderate even in the greatest waves. Hence wind can speed up waves that are moving faster than itself, although this speeding up is opposed by the wind resistance of the wave crests. From measurements it appears that the two tendencies balance when the wave speed becomes 1.45 times that of the wind. Beyond this there is no increase in wave speed, no matter how long the wind may blow.

Thus wave size and speed depend on several factors: wind speed, wind duration, the *fetch* (the distance across which the wind blows with the same direction and speed) and, finally, the state of the sea at the time the wind began to blow. Fairly good formulas involving these factors have been worked out to predict the height, speed, and period of waves. These were used with remarkable success in prophesying surf conditions for the Normandy Invasion and other amphibious operations during World War II.

Waves generated in an area of wind may progress into a region of calm or of different wind direction. Here they are called "the swell." They lose height and energy because of air resistance to the wave fronts and internal friction. Somewhat surprisingly, they also become longer and speed up. Roughly, the swell loses one-third its height in traveling a distance in sea miles equal to its wave length measured in feet. Nevertheless, some of the greatest waves that pound the coast of southwest England have traveled thousands of miles from their source area in the South Atlantic.

When either "forced" waves, which are still growing with the wind, or "free" waves, like the swell, which are no longer growing, enter shallow water, several things happen. When the depth becomes less than L/4, the bottom interferes with the orbital motion of the water particles, causing the wave to slow and steepen. Where the water shallows to L/20, the wave speed no longer depends on L but becomes proportional to "d Such waves are called *shallow-water waves,* as distinguished from the deep-water waves whose speed is independent of the depth.

WAVE REFRACTION

A deep-water wave thus begins to "feel bottom" where the depth is about a quarter or, at most, half the wave length. Commonly, this is less than 500 feet, but for the largest storms it may be nearly three times this depth. Though the wave period remains the same, the wave length shortens—in other words, the wave slows down.

This is not because of friction but because the water particles at the bottom can no longer move in circles; their paths become ellipses, and directly at the bottom are reduced to straight lines, to and fro, in the direction of wave advance. Where the wave approaches shore obliquely, the inshore end feels bottom sooner than the offshore parts. It therefore slows sooner and the wave becomes bent, for the offshore part continues to advance as a deep-water wave (with greater speed) after the inshore end has begun to change to a shallow-water wave. Along a straight, evenly sloping coast, then, no matter what the original angle of approach may have been, the waves tend to be bent (refracted), so that they approach shore nearly straight on—that is, with their crests nearly parallel to the beach.

Wave refraction modifies the distribution of the energy in the waves. The energy which was uniformly distributed along the wave length in the deepwater wave becomes concentrated on the headlands and diminished in the coves. Hence the waves on the headlands are more powerful and higher, those on the shores of the coves less powerful and lower, than they would be on a straight coast. We shall see that this selective attack on the headlands makes the waves tend to shorten and straighten the shoreline by eroding the headlands and filling the coves with the transported debris.

Breakers

In deep-water waves only half the energy is carried forward with the speed of the wave. The other half is used in the oscillation of the water particles above and below their quiet-water positions. On advancing into shallow water little of the energy is used up in friction; the wave is, nevertheless, slowed up by interference with orbital motion near the bottom and, since its energy is nearly the same, its height must theoretically increase. On the other hand, when the wave has finally been transformed to a shallow-water wave, all of its energy is being carried forward with the wave. Theoretically, this should lower the wave height, since a smaller mass of water can carry the same energy as the larger mass of a deep-water wave, which is only 50 per cent efficient in transporting energy. These theoretical conclusions are borne out by observations of real waves.

Waves whose ratio of height to length in deep water is more than 1/100 steepen on entering shallow water until the crest angle is about 120°. When they are this steep they become unstable and begin to break but may continue for hundreds of feet without curling over of the crest. This is the ideal wave for the surf-board rider—it is often only the exceptionally high wave that will do this. On the other hand, waves in deep water that are lower relative to length carry most of their energy in the wave crest. As these slow and steepen in shallow water they no longer find water available to fill their wave forms. At this point the wave curls over, breaks with a single great crash, and the water tumbles forward onto the beach. This is the breaker. Depending on the wave period, breakers occur where the still-water depth is from one to two times the wave height.

When a wave breaks, the energy it carried is dissipated in collision with the bottom and in turbulence of the water. Bottom materials in the breaker zone are thrown violently into suspension, as any surf swimmer can testify. Because of the concentrated turbulence, a trough is often scoured in the bottom along the breaker line. Material thrown into suspension sweeps forward with the rush of the water up the beach. As the water shallows and becomes less turbulent, the detritus is sorted according to its size, shape, and density.

WAVE EROSION

Waves erode the shores in three ways: by their impact and *hydraulic pressure,* by *corrasion*—the sawing and grinding action of the sand, gravel, and cobbles hurled against the cliffs or rolled and dragged across the foreshore —and by *solution,* which is a minor process even along limestone shores.

Hydraulic Action

Waves contain astounding energy. Even moderate waves, 10 feet high and 100 feet long, exert a pressure (theoretical) of 1,675 pounds per square foot when they strike an obstruction. Great storm waves, 42 feet high and 500 feet long, should produce a pressure of more than 3 tons per square foot. On the east coast of Scotland pressures as great as this have actually been measured by dynamometers.

At Wick, in the north of Scotland, waves of a great storm

in 1872 tore a concrete monolith (45 feet long, 26 feet wide, 11 feet thick, and bound by 3½-inch iron rods to the breakwater foundation) from its place, together with a great mass of the foundation, and dropped the whole unbroken inside the harbor. The mass weighed 1,350 tons.

The block was replaced by a much larger mass of concrete weighing 2,600 tons; five years later this, too, was carried away by the waves. At Tillamook Rock lighthouse on the Oregon Coast, waves have repeatedly hurled stones large enough to break the plate glass of the light 1.32 feet above the sea. One fragment, weighing 135 pounds, was thrown through a roof more than 100 feet above sea level, breaking a hole 20 feet square. At Dunnet Head, north Scotland, the lighthouse windows 300 feet above the water are occasionally broken by stormdriven stones.

The tremendous wave pressures act not only directly, but also by compressing air driven into crevices in the rocks. Such pressures suffice to pry huge blocks from cliffs exposed to full-wave attack. At Ymuiden, Holland, a 7-ton block in the breakwater was seen to move forward—toward the sea — presumably because of compression of the air in crevices behind it.

Thus storm waves seek out weak strata or joint cracks in a cliff and dislodge the blocks between them. Sea caves so formed may be scores or even hundreds of feet long; some penetrate entirely through small promontories and produce spectacular arches. The roofs of others collapse, leaving isolated *stacks* in front of the cliff line.

Corrasion

As we have seen, great storm waves can tear huge blocks from the cliffs and hurl them against the shore. Even moderate waves can move boulders and cobbles, and feeble ones sand. In a Cornish mine that extends beneath the sea, it is reported that the grinding of boulders is easily heard through a 9-foot roof of rock, as the waves beat against the cliff overhead. The shore zone is a veritable grinding mill. On the beach at Cape Ann, Massachusetts, angular fragments from the granite quarries have become rounded within a single year. The sea is a great saw; armed with cobbles, pebbles, and sand, it works away at the beach, undercutting cliffs and wearing a notch in every rock exposed at this level.

In relatively unconsolidated material, such as the glacial gravels making up the Holderness coast of Yorkshire, the sea has eaten back the cliffs at a measured rate of 7 to 15 feet per year for more than a century. Shakespeare's Cliff, several hundred feet high, a part of the "Chalk Cliffs of Dover," becomes undercut so as to produce frequent large landslides. One of these, in 1810, was so great that the vibrations were felt as a strong earthquake at Dover, several miles away. These are spectacular examples, but even in hard rocks, like the rhyolite, wave-notching is impressive.

Corrasion is not confined to the water's edge. By far the greatest energy is spent between the breaker line and the shore; nevertheless, particle motion at greater depth is considerable, especially if the length of the waves is great. Waves with a period of ten seconds that are 500 feet long and 22 feet high should, theoretically, produce water particle speeds of 10 inches per second at 300 feet depth; this is enough to move fine sand.

When account is taken of tidal and other currents, it appears that the bottom water is commonly agitated enough to keep clay particles in motion to depths of 600 feet and on some coasts to 900 feet or even more. These conclusions from wave theory are partly confirmed from the character of the sediments found by dredging. The depth to which wave motion is perceptible is called *wave base*.

Now, it is true that the particle speeds mentioned reverse their direction with every wave. But most of the shallow sea floor slopes seaward, so that a particle moving to and fro on the bottom is moving downhill seaward and uphill landward; outside the breaker zone, where these oscillations are about equal, the sediment tends ultimately to progress downhill (seaward) in response to the pull of gravity.

As the motion becomes less and less with greater depth, the coarser particles eventually reach positions where agitation is too feeble to keep them moving. Here they come to rest. But finer particles can still be moved and will continue to travel seaward until they, in turn, are sorted out in accordance with their size.

During all this agitation the particles of sediment wear against each other and against the bottom. They gradually become smaller and smaller until they can be moved by successively feebler oscillations. Dragged back and forth with each passing wave, they also wear away the bottom, so that not only the breaker zone, but all the coastal zone to the depth of wave base, is corraded.

Inshore from the breaker zone, where the waves have become forwardmoving masses of water, the situation is somewhat different. The breakers dashing onshore may carry coarse fragments to heights where the feebler backwash cannot carry them back, even though it is favored by a steep slope. Although as much water must flow away from shore as toward it, the shoreward movement is generally concentrated in a shorter time, so that the velocities in the breaking waves are greater than in the flow of the backwash. A coarse pebble or cobble *storm beach* may thus be built above the reach of the normal waves. Such a beach is steep, for coarse material can be moved seaward only on steep slopes.

Eventually, a profile is developed for any particular size of waves at which swash and backwash are about equally effective. Even after this profile is developed, the beach material is gradually ground down in size so that slow seaward transport continues. As the waves fluctuate in size, the profile and the grain size of the material continually readjust to conform to them.

THE PROFILE OF EQUILIBRIUM

Thus, partly from wave theory, partly from observing the movements of detritus in waves, partly from dredgings and soundings, and partly from study of large, dried-up lakes the concept of the *marine profile of equilibrium* has been developed. This profile is a smooth, sweeping curve, concave upward. It is steep in the breaker zone and flattens seaward. Its slope is adjusted to the average particle size of the detritus moved by the waves at each particular place along it. The slope is that which just suffices to keep the material in slow transit toward deep water. As the waves forever change in size, the profile constantly readjusts, flattening and shallowing with feeble waves, steepening and deepening with more powerful ones. But for any coast there should be, theoretically, a general average slope about which the profile fluctuates.

The concept of the profile of marine equilibrium is not purely hypothetical. Soundings commonly show that the sea floor near shore is concave upward, despite the complications to be expected because of the world-wide changes in sea level that occurred during the Pleistocene. The slope and depth of the offshore profile is more or less closely adjusted to the fetch and power of the waves. The reality of such a profile-at least of its shoreward portions—is still more convincingly shown by examination of uplifted sea floors in many places. Such graded sea floors, uplifted to form *marine terraces*, are found in many places. They show some or all of the features seen and deduced from modern coasts: cliffs and cliffnotches, stacks and caves, storm beaches, gravel and sand beaches, and a sedimentary veneer resting on a smoothly truncated surface of bedrock whose profile is gently concave upward but retains a general seaward slope.

It is convenient, then, in geology, to refer to the profile of equilibrium shore processes. But it should be clearly understood that the idealized profile is rarely achieved, partly because of the continual changes in wave power and the complicating effects of tidal and other currents, and probably mainly because of the fluctuations in sea level and slow movements of the earth's crust that have interrupted its development. Nevertheless, profiles approaching the ideal are common, and beach materials can be seen to move in sensitive response to natural and artificial changes in the profiles.

Tides

The ancients knew that the ebb and flow of the tides varied with the phases of the moon. So complex is the real earth as compared with the idealized earth assumed by the astronomers and physicists that we have, as yet, no general theory that permits tidal forecasts for any point on an ocean. Tides are, of course, predicted with great accuracy for all the principal ports; these are not computed from general theory, however, but from analysis of tidal records over a long period of years at the particular port concerned.

The sun's attraction acts like that of the moon. Despite its far greater mass, it is so much farther away that its maximum tide-producing force is only 0.46 that of the moon. Twice during the lunar month, at new and full moon, both sun and moon lie on a straight line that also passes through the earth and their influence is greatest. These are times of extreme tides (*spring tides*); at all other times their tide-producing forces tend in some degree to neutralize each other. At the first and third quarters their influences are directly opposed and tides are smallest (*neap tides*). The essentials of the tide-producing forces, it certainly fails to explain fully the vagaries of the local tides. For example, many ports have but one tide in a lunar day; in others the lag is many hours from the time the moon is at zenith; in still others the two daily tides are of greatly different heights. They also vary with the seasons.

These and many other facts make it clear that the tides are not a simple direct response to the vertical component of the moon's gravitational pull, which is really far too small for effective lifting of the water masses anyway. Tides are really great shallow-water waves, induced by periodically shifting tidal forces acting on the water particles of the sea in much the same way that wind turbulence produces water waves. And, like other shallow-water waves, they "feel bottom" and are influenced by its configuration. Depending on this configuration, and on still other factors only little understood, they produce currents that vary greatly in strength and direction at different points and at the same point in different seasons.

To the geologist the principal interest of the tides is in these currents. Some are prodigious, particularly in estuaries with converging shores. In the Bay of Fundy, between New Brunswick and Nova Scotia, the extreme tidal range is as much as 70 feet vertically, and the possibility of harnessing the power of the huge water masses involved in this vertical movement has been seriously studied. Thus far, the great cost of the dams needed and the fluctuation in amount of power available from hour to hour and day to day have made the project seem uneconomic.

The tidal currents attain speeds of 9 miles per hour during both rise and fall of the tide. In places they have scoured basins on the bottom more than 150 feet deep. In St. Malo Bay, Brittany, the tidal range is 40 feet, and currents reach speeds of 8 miles per hour. Between the Orkney and Shetland Islands the tidal currents move at 12 miles per hour. In some rivers the tide enters as a high wave of water called the *bore*. The bore of the Hangchow River in China is as much as 16 feet high and speeds upstream at 16 miles per hour.

Since experiments have shown that water moving at ½ mile per hour will transport medium sand grains, and at 3 miles per hour gravel an inch in diameter, it is clear that tidal currents are important agents of erosion and transport of sediments. Dredgings off the Mull of Galloway in Scotland show that coarse gravel is moved by tidal currents at depths of more than 800 feet.

Because the wave length of the tide is so great, it always acts as a shallowwater wave—that is, the greatest ocean depth is only a small fraction of the wave length. This is, of course, oversimplified). Tidal waves must everywhere reach bottom and produce some water motion even at great depths.

Because of the density stratification of the sea, secondary internal waves are formed at the boundaries of water masses of different density, and bottom motion is probably negligible in many places. The currents should be strongest near the edges of the shallow continental shelves because, here, more water must flow in and out (in proportion to depth) than elsewhere. This may account, at least in part, for the sediments near the outer edge of the shelf being somewhat coarser than those farther inland. The stronger currents should winnow out the finer sediment and drop it off the shelf onto the continental slope.

Over submarine ridges, regardless of depth, tidal currents must be perceptible where the ridge rises steeply from the bottom. Many ridges yield bare rock to the dredge, and nearly all show coarser sediments than adjoining basins or depressions.

In summary, we have reason to think that the winnowing action of tidal currents significantly affects the size, sorting, and distribution of sediment on the sea floor. But only locally are they important in shaping the shore itself and, in general, their effects are subordinate to those of wave currents. Two wholly different kinds of waves, neither of them tidal, have, unfortunately, become popularly known as "tidal waves." One, the sea wave caused by earthquake, *seismic sea wave* the other is a high-water wave caused by prolonged and unusually violent onshore winds.

Wave Currents

Despite refraction, most waves approach shore at a slight angle, so that, on breaking, the swash has a component of motion parallel to the shore. This produces a current called the "longshore drift," which, depending on the exposure of the shore, may consistently flow in one direction or reverse with differences in the wave approach. Such a drift may become swift, especially if there is a longshore wind, or if tidal currents reinforce it. The breaking wave carries sand and gravel obliquely up the beach—the sand particles farther than the gravel because they move with a feebler current—and even the backwash may retain a longshore component. Each particle thus moves in a series of sawtoothlike oscillations down the beach. Marked pebbles made from bricks have been traced along the upper beach as much as half a mile in a single day; similar movements doubtless occur in the whole bottom zone agitated by the waves. Vast amounts of sand are transported parallel to the shore by these longshore currents.

ARTIFICIAL INTERFERENCE WITH SHORE PROCESSES

Artificial interference with waves and currents furnish

many examples of the nice adjustment of beach profiles and shore outlines to the average power of waves and currents. Even a slight obstacle to movement of detritus often causes drastic changes in the shape of the beach.

Dredging of sand and gravel offshore has frequently been followed by erosion of the shore; the material left behind the excavation is swept seaward to restore the slope that had been modified by removal of the dredged material. Still more striking results have come from the building of groynes on many bathing beaches. Groynes are low walls extending from the high-tide line out into the sea with the purpose of preventing the removal of sand by longshore drift. Where they have been successful in holding the fine sand, the downdrift beaches, being deprived of their normal supply of traveling sand, have often been severely scoured and changed from sand to gravel or cobble beaches.

Breakwaters, walls of rock or concrete built to protect anchorages from storms, have often greatly modified beaches, as at Santa Monica, California. A 2,000-foot breakwater was built parallel to the shore and about 2,000 feet away from it, in order to furnish anchorage for small boats. The general longshore drift here is to the southeast. The breakwater, of course, reduced the power of the waves striking the shore behind it. The detritus moving along the beach was no longer kept in suspension in the shore zone, and the beach in the lee of the breakwater began to fill in. In less than 20 years it had advanced more than 500 feet.

Despite being robbed of their normal quota of drifting sand from the northwest, the beaches to the southeast of the breakwater have not been eroded inland very far, although it is noteworthy that in time of great storms there is considerably more damage to the southeast of the breakwater than to other nearby parts of the shore.

EROSION AND DEPOSITION BY THE OCEANS

The Shores of Mountainous Coasts

The Shores of mountainous or hilly coasts are more varied

than those of plains. Waves attack exposed promontories, producing cliffs; the coarse and fine debris eroded from them is distributed by longshore drift to make characteristic shore features—bay-head and bay-mouth bars, land-tied islands, spits, and bars.

The wave refraction on the shores of a bay produce currents that tend to sweep material to the head of the bay, where it accumulates as a bay-head bar.

Material carried by longshore drift tends to continue straight across indentations in the coast. Debris from headlands may thus tend to close a bay by forming a sand spit that continues in a smooth curve from a cliffed point. Toronto Harbor is protected by a long sand spit growing westward. As such spits are built out into deep water, wave refraction and, in the sea, local tidal currents, cause their ends to become curved. Where the outflow from a bay is small, such spits may build completely across the bay, forming a bay-mouth bar. The lagoon behind may be brackish or fresh and may be slowly filled by silt from the land or by sand blown inland from the beach.

Islands refract the waves and protect the beach behind them from the full force of the sea. Hence the longshore drift cannot effectively carry material behind an island. Therefore the mainland beach tends to be built out, diverting the longshore currents toward the island. Eventually, a spit may extend to the island, tying it to the shore. Many islands thus become joined to the coast in both directions, enclosing triangular lagoons.

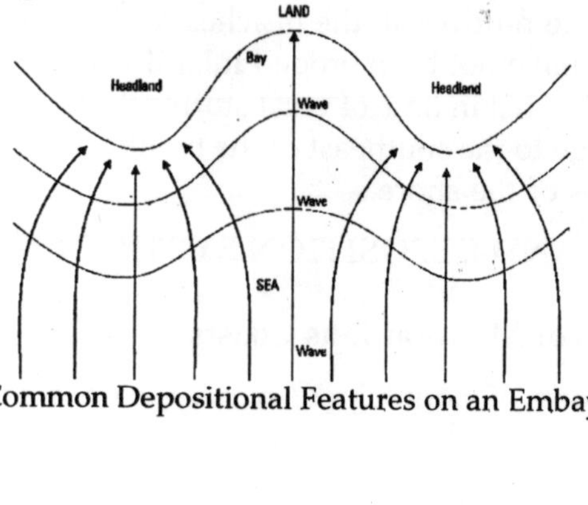

Fig. Two Common Depositional Features on an Embayed Coast.

Even on straight coasts the drift may be deflected by bottom irregularities and build forward into the sea. The beach formed varies in height with the detritus supplied and the height of the storm waves. As the beach advances seaward it leaves behind a series of older beach ridges—some 20 feet or more high—separated by swales. Perhaps the best known of these advancing beaches is that of the Dungeness, in southeastern England, which has grown more than a mile into the sea since the time of Elizabeth—about 6 yards per year.

The Shores of Plains

Along low-lying coasts such as those of the Gulf of Mexico and the southeastern Atlantic states, the wave attack is spread widely instead of being concentrated upon headlands. The sea deepens gradually offshore, hence the waves feel bottom far out, and the material they stir up is built into an *offshore bar* just inland from the zone of greatest breakers. As wave agitation must continually grind down the bar material and slowly move it seaward, offshore bars must advance slowly landward in order to survive with a stable sea level. Most offshore bars along plains do touch the mainland locally. Here the shore is undercut to produce low cliffs. Part of the material eroded is carried by longshore currents and added to the bar.

The lagoon behind an offshore bar is, of course, salty, but rivers may freshen the water considerably. A large river may raise the lagoon level to such a height, especially at low tide, that the bar is broken, forming an inlet. Tidal currents help to keep the inlet open. Inlets shift from time to time because of shore drift, and new ones are sometimes formed. In the protected lagoon, salt grass grows. Gradually the lagoon changes to a marsh as plant debris accumulates along with the silt brought by streams, tidal currents, and winds.

The famous Florida and New Jersey beaches are such offshore bars. Another striking example is the long sand bar that fringes the southern coast of the Baltic from Danzig to Memel, enclosing an almost continuous brackish lagoon. A plains shoreline along which some offshore bars still enclose lagoons, though others crowd meandering tidal rivers behind them. Magnificent examples of offshore bars fringe the Carolina coast. Their cuspate shape has been attributed to the junction of eddies thrown off by the Gulf Stream but it may be partly controlled by wave refraction from bottom irregularities offshore.

Beaches that dry out between tides furnish sand which may be blown inland by winds. Because of the absence of coarse shingle and the prevalence of sand beaches along plains coasts, the country inland is likely to be overrun by sand dunes wherever onshore winds are common.

SEDIMENTS OF THE SEA

The Sources of Marine Sediments

Except for a trivial amount of meteoritic dust and the much greater increment from submarine volcanoes, all marine sediments come ultimately from the land. Even the material of marine shells has been leached from rocks of the lands, though it is extracted from solution in the sea. Sluggish or swift, the rivers sweep the land waste seaward. Detritus from rivers and rills, from downslope movement, from glacial plucking and abrasion, along with windblown volcanic ash and desert dust, are carried to the sea, there to mingle with the detritus won by the ocean itself in its attack upon its shores.

The rivers bring to the oceans about twenty billion tons of rock waste every year. More than four-fifths of this is elastic material carried in suspension or rolled along the stream bottoms; the rest is dissolved. The fate of the elastic material is clear. Samples dredged from the sea floor show partly consolidated sediments identical in grain size and mineral content with river sediments.

Stratification is more common, though not universal, and sorting more complete, than in sediments deposited by rivers. Furthermore, the widespread marine sedimentary rocks that are uplifted to form part of the lands show that the solid particles carried to the sea by streams accumulate on the bottom. The fate of the dissolved material is not so obvious. We know that some of it is deposited, for there are great marine beds of limestone, dolomite, salt, and anhydrite, but we do not know whether deposition exactly keeps pace with replenishment and, hence, whether the sea is becoming more or less salty.

As was previously noted, river waters have a much larger proportion of silica and calcium than sea water, so that if these substances were not withdrawn from solution in the sea, its composition would change greatly within a short time, geologically. Life processes regulate this; limy and siliceous shells of various organisms represent much of the material brought to the sea in solution.

Marine organisms are geologically important not only because they contribute their shells to sediments but also as indicators of the environment of deposition. An understanding of the main features of marine life is essential for interpreting the sedimentary rocks and the historical record they contain.

The Life Zones of the Sea

Life is everywhere in the sea. Even the greatest depths—far below levels to which light penetrates, and where the water temperature is only a little above freezing—yield living organisms to the trawl. Although marine environments are less variable than those on land, they, and the organisms living within them, do differ from place to place. Marine biologists classify the different parts of the sea.

The *two* main divisions are the *benthic,* or sea-bottom environment, and the *pelagic,* or open-water environment. The benthic environment is divided, at a depth of 200 meters—about the depth of the edge of the continental shelf—into a *littoral* system above and the *deep-sea* system below. The depth of 200 meters is critical because it is near the limit to which light can penetrate. On the deep-sea floor there is neither light nor seasons, and hence no photosynthesis by green plants. The bottom life is confined to scavengers living on organisms sinking from above and to a few bacteria, nearly all of which depend on organic compounds formed ultimately by photosynthesis but brought to this depth by currents or by sinking.

The life of the sea, both plant and animal, is classed in three large groups: the *benthos,* or bottom dwellers; the nekton or swimming forms; and the *plankton* or floating and drifting organisms. Of these, the benthos includes all the attached, creeping or burrowing organisms of the bottom: seaweeds and grasses, sponges, barnacles, clams, oysters, corals, bryozoa, worms, lobsters, crabs, and many other animals. Among the inconspicuous but geologically important organisms of the benthos are the minute single-celled animals called *Foraminifera,* most of which secrete shells of calcium carbonate. The nekton includes the squids, fishes, seals, whales, and many other animals which are economically important but of little geological significance.

The plankton includes all the surprisingly varied organisms, chiefly microscopic, that float with the ocean currents. Among them are several kinds of diatoms with shells of opal, the *Coccolithophores,* a group of algae that secrete rounded shells made up of calcite in thin plates, many individual *Foraminifera* and the silica-secreting Radiolaria. As many as 100,000 diatoms have been found in a quart of water from the open North Pacific, though the usual abundance is far less.

Except for a few very specialized bacteria, all life depends on photosynthesis—the process by which plants utilize radiant energy from the sun to make compounds of carbon, hydrogen, and oxygen. These compounds constitute the bulk of the plants, and are used directly or indirectly by animals that eat the plants or other animals. Marine plants, like land plants, form organic compounds directly from carbon dioxide and water, but only in the presence of light. Plants require small amounts of many substances, among them phosphates, nitrogen, iron, and manganese. These factors of food supply that limit plant growth are partly reflected in the very different densities of population in different parts of the sea, and in the comparable variations in the animal life that feeds on them.

When an organism dies, unless it is eaten by another animal, the organic compounds comprising it ultimately sink below the lighted zone, perhaps to feed some scavenger in the depths. In sinking, much of the organic matter is decomposed into carbon dioxide and other constituents by bacteria. This decomposition both uses up the oxygen dissolved in the water and adds to the carbon dioxide in it—two factors that strongly influence the kinds of sediments of the deep sea.

Decomposition at any depth returns the phosphorus, nitrogen, and other nutrient elements to solution as simple compounds once more available for plant growth, if the water mass containing them returns to the lighted zone. These factors account for the flourishing plant life in parts of the sea and explain why the Newfoundland Banks are such prolific fishing grounds, as are similar areas of upwelling along the west coasts of California, Chile, and Africa.

REALMS OF SEDIMENTATION IN THE SEA

The "Challenger" expedition proved that sediments are being deposited virtually everywhere in the sea except where currents scour the bottom or where slopes are too steep to retain them. Even where depth, exposure, distance to shore, submarine slope, and other obvious factors seem identical, the local sediments vary greatly. Despite much careful work, we are far from understanding all the controlling factors. Sir John Murray, the leader of the "Challenger" expedition, classified the sediments into two great groups: *terrigenous* (derived from land), and *pelagic* (of the sea). The *red clays*, some of the most representative of his pelagic sediments, are actually composed largely of land-derived material.

Also, the bottom mud at a depth of 33,000 feet in the Mindanao Trough is indistinguishable from many continental shelf deposits; it is a "terrigenous mud." Despite these anomalies, deposits of different realms vary broadly and some of them sharply. Modern oceanographers call sediments "pelagic" if they are thoroughly oxidized—a condition that indicates long contact with deep water and hence slow deposition. Deposits of any depth that show lack of this oxidation, and hence accumulation rapid enough to prevent it, are called "terrigenous" even though found at greater depth.

We accordingly consider the *terrigenous sediments of the*

strand zone, the *terrigenous sediments of the shelf zone*, and the *pelagic sediments* as different classes. Because of their geologic importance and the peculiar problems they offer, we shall consider separately the deposition of carbonate rocks (limestones and dolomites).

Sediments of the Strand Zone

The strand zone is the area immediately along the shore. Its sediments differ considerably from place to place. The most important kinds are beach deposits, deltaic deposits, and tidal-flat deposits.

Beaches

Beaches are deposits of cobbles, gravel, and sand on and near the shore. During great storms much or all of this material is moved about or even stripped away by the waves, but usually only the superficial layers are agitated. Beaches vary greatly in size; some between cliffs and headlands are only a few feet long; others on plains coasts stretch uninterruptedly for miles. They range from a few feet to several thousands of feet in width. Most beaches of coarse material are narrow because steep slopes are necessary for coarse material to move seaward. Even in sand beaches the steeper ones tend to be made of coarser sand than the flatter ones, though the correlation of slope and grain size is not perfect. Most beaches are concave upward because the coarsest material is nearest the shore.

The waves not only wear down and round the grains, but also sort them according to size. Variations in the waves are reflected in the size of grains they can move. Thus the beach develops a laminated structure. The perfection of sorting and lamination depends on the available material. It is much poorer along coasts of high relief (which generally supply heterogeneous material) than along plains coasts (where the source material has already been sorted by sluggish streams or the waves of a former sea).

Beach sands may be largely of quartz if they come from deeply weathered granitic regions or from reworking of sedimentary rocks, for such reworking means, of course, that any feldspar originally accompanying the quartz will have been exposed to weathering at least twice. But in regions of rapid erosion or of ineffective chemical weathering, feldspar abounds on sand beaches.

In Hawaii, whole beaches are made up of grains of olivine from nearby basalts. Some beaches, particularly in the tropics, are composed largely of shell fragments. Along the Brazilian coast these are cemented almost as fast as they are being formed because the spray dissolves a little calcite at the surface and redeposits it below. Many sand beaches contain concentrations of heavy minerals ("black sand"), such as magnetite, chromite, and garnet, sorted out by the wash because of their high density as compared with other minerals. At Nome, Alaska, gold thus concentrated has long been worked.

Few beach deposits have been recognized among the rocks, though recently uplifted beaches are abundant. Beaches are quickly destroyed because uplift exposes them to erosion, whereas submergence to a depth less than wave base subjects them to reworking by waves and tidal currents. Nevertheless, Cambrian beaches, with associated cliffs and tumbled stacks, have been recognized in the Montana Rockies, Carboniferous and Jurassic beaches in Great Britain, and other examples elsewhere.

Deltas Streams generally form deltas where they enter the sea, but if the stream load is small or the coast exposed, the detritus may be widely scattered by waves and currents, and no delta is formed. Deltas furnish another example of the difficulty of geologic classifications, for all ocean deltas contain some beds that are truly marine, others that are brackish lagoon deposits, and still others that are fluviatile—all complexly interfingering.

Where a stream enters the sea the surface slope that gives it a current vanishes. If wave action is strong, river water and sea water quickly mix, but along quiet coasts the fresh water may float for awhile above the salt water. The powerful flood of the Amazon extends far into the Atlantic, and fresh water may be dipped from the surface scores of miles out to sea. The sediment suspended in the river water is thus distributed widely over the sea floor. The water of some rivers is so heavily laden with sediment that it sinks under the ocean water and flows down the slope of the sea bottom. Ultimately, of course, such sediment-laden water will deposit its load, either in a submarine delta or widely over the sea floor. There is growing evidence that similar sediment-laden "density currents" may form independently of rivers within the sea itself.

Rivers vary widely in the grain size of their sediment and all carry coarser material in flood than in low water. Some, like the Colorado, carry much coarse sand; others, like the Mississippi and the Rhine, discharge little but clay, silt, and fine sand. At the river mouth, the coarsest material settles out of suspension first, followed progressively by finer and finer grains, as can be seen in the Mississippi delta where coarser silt and fine sand extend on either side of the distributary channels out into the Gulf. Traced landward, these submarine ridges emerge and become natural levees. Like all natural levees, they are made of coarser sediment than is found behind them.

The activities of delta distributaries vary with river discharge and with shifts to other newly formed distributaries. Thus some segments may advance into the sea while other segments are being destroyed by the waves. The shore processes are like those of a plains coast. Offshore bars sheltering tidal lagoons are common; the waves may push the bars inland to the delta front and then rework the stream-laid beds, concentrating the coarser material on the beach and strewing the finer over the sea floor.

The factors just outlined condition deltaic sediments; the coarser material comes to rest chiefly on the stream bed or on beaches; the next finer forms natural levees and their submarine extensions; the finest sediment is deposited in back of the levees during floods or is swept out to sea.

The clay particles in river water have very large surface areas relative to their bulk. They are dispersed as individual particles, largely because the unsatisfied valence bonds of the surface ions attract negatively charged ions such as O^{--} and OH^{-} and thus prevent one particle from coming in contact with another. On entering the sea, with its abundant ions of Na^{+} and Mg^{++}, the surface shells of attracted ions are stripped away, and the clay particles come into contact and adhere, forming larger aggregates which quickly sink.

Clay should, therefore, settle quickly when it enters the sea. Actually, few careful studies of this process have yet been made, though the above idea conforms with laboratory experiments. In one study off the California coast, Revelle and Shepard found that, despite thorough mixing of stream and ocean water, much clay was carried by eddies and currents as far as 150 miles offshore before settling to the bottom, though most was deposited much nearer the coast.

Some deltas, like that of the Mississippi, are known from surveys to be subsiding, locally as much as 8 feet per century. It has been suggested that this is due to the load of sediments added to the crust. As with glacial loading, this seems a necessary consequence of isostasy, although the sinking of the Mississippi delta seems too great to be explained in this way alone. The unconsolidated sediments are much less dense than deeper-lying rocks and hence should not cause their floor to sink as much as their own thickness. But, whatever the cause may be, parts of the subaerial delta are carried below the sea, forming tidal lagoons or even open channels where fine marine sediments accumulate.

Perhaps this subsidence gives the Mississippi delta its peculiar bird-foot form by submerging all the outer surface except the natural levees. As distributaries shift, deltaic marine beds may be buried by river deposits. If subsidence continues, the delta may remain nearly in place and not advance into the sea. Thus interlayered marine, fluviatile, and brackish-water deposits may accumulate to great thickness, thinning into beds completely fluviatile landward and marine seaward. Many deltas have been recognized among the ancient sediments by the intimate association of these varied deposits. In New York and Pennsylvania a delta formed during the Devonian shows strata more than 15,000 feet thick; borings for petroleum indicate a comparable thickness of the Mississippi delta along the Louisiana coast.

Tidal-Flat Deposits

Tidal-flat deposits accumulate in estuaries of sluggish streams or behind offshore bars. Many fill quickly; at Wilhelmshaven harbor and Cuxhaven, both on the North Sea coast of Germany, 8 to 10 feet of mud accumulates in a single year. Much of this mud and silt is brought by rivers, the rest must be swept in by the tide. Some of the mud from the German and Dutch coasts is so full of organic matter that it is used as agricultural fertilizer. Shallow parts of tidal lagoons commonly support a luxuriant growth of brackish-water plants. This vegetation speeds up the silting of the lagoon, for it acts as a filter to screen out sediment.

Like beaches, lagoonal deposits are only exceptionally preserved among the older rocks. Nevertheless, some of the Devonian shales of Germany, as well as other examples, have so many of the characters of modern lagoonal deposits that they are thought to represent this environment.

Sediments of the Shelf Zone

The "profile of equilibrium" is an ideal concept. Soundings and samplings of the continental shelf sediments seem to show actually as much deviation from a smooth profile as agreement with it. Nor is there a notable correlation of grain size of sediments with either distance from shore or depth of water. It does appear that the immediate shore zone generally has both a flattening profile and finer sediments seaward, but within a short distance from the shore both these regularities usually disappear.

Such deviations may be partly explained by the rise in sea level of perhaps 300 feet since melting of the Pleistocene glaciers—a time so short that the sea has not yet had time to adjust its submarine profile and the grain size of its sediments to its new level. Or perhaps crustal movements have similarly interfered with the development of a smooth slope and effective sorting. It seems, though, more likely that the currents set in motion by the winds, waves, and tides act in a more complex way than the idealized theory envisages. Whatever the reason, the fact remains that many recent studies have shown that the simple graded slope, steeper inshore and flatter offshore, and covered with sediments that become regularly finer seaward, is far from characteristic of many of our present seas.

SEDIMENTS OF THE ATLANTIC CONTINENTAL SHELF

The material reported is largely that adhering to the tallow on a sounding lead. The charts show that, far from having a simple gradation from coarse to fine seaward, the sediments are very irregularly distributed. There is actually more sand nearshore and more gravel in deep water. The sediments of the outer edge of the shelf are coarser than those inshore. Silt and clay, however, are rare at depths less than 600 feet and extend from here down the continental slope.

The coarser materials within 20 miles of the shore are very irregularly arranged. This profile has been developed on an area glaciated during the Pleistocene, and the advancing sea has, therefore, had material of all sizes from boulders to rock flour available for sorting. Profile B, off New Jersey, where the floor was not glaciated, shows finer material inshore but considerably coarser material at the same depth as that at which Profile A shows silt.

Here, too, grain sizes of sediment are irregularly distributed, but beyond 25 miles from the coast the tendency is toward finer material seaward except near the shelf edge, where, again, coarser material prevails. Silt may be rare along this profile because the sea is here working on old coastal plain sediments that contain little but sandsized material. Profile C, off northern Florida, is very regular in slope, but even here the material outside the 25-mile limit is consistently coarser than that inshore. Strong currents of the Gulf Stream sweep the outer edge of this profile. The coarseness of sediment here shows clearly that the Gulf Stream is swift enough to carry fine sediment right across the shelf, down the slope, and even across the Blake Plateau, whose outer edge is marked by the 2,000-meter (6,500-foot) depth line. It is concluded that nearby a fairly coarse sand is being deposited at depths greater than 6,500 feet.

SEDIMENTS OF THE DUTCH EAST INDIES

The submarine topography of the Dutch East Indies is highly irregular, with deep basins, shallow marine plains, and submerged mountain ridges. Some of the basins—the Sulu, Celebes, and Banda seas—are more than three miles deep, separated from the ocean basins by much shallower ridges and saddles. The situation is further complicated by the many volcanoes, of which more than 50 have been active at some time since 1600. Although only a few hundred samples of the sediments of this great region of more than a million square miles have been studied, some extremely interesting results have been gained, chiefly by the Netherlands "Snellius" Expedition of 1929-30.

When Tamboro, on an island east of Java, erupted violently in 1815 the ash was thrown high in the air and carried by the winds for tremendous distances. This ash is readily recognizable from its distinctive minerals. In places 240 miles distant from the volcano as much as 10 inches of this ash was deposited.

Even more striking than the prodigious volume of the ash is its presence as the topmost layer in submarine cores collected from the basins of the region. Although it has been suggested that some of the very finest grains of this ash are actually still being deposited—reaching bottom for the first time 115 years after the eruption—it seems more likely that reworking of the sediment by benthonic animals may account in part for its abundance in the surface layer.

Rock bottom was found on many of the ridges between the East Indian basins at depths of several thousand feet. Fine muds were found at far shallower depths in less-exposed places; thus we have evidence of strong currents sweeping the bottom clear of sediments at depths far below the zone of agitation by surface waves. Land-derived mud was cored fully

300 miles from the nearest land and at depths of 16,000 feet; volcanic muds from still greater distances and depths. From the Philippine Trench, east of Mindanao, 50 miles from land and at a depth of 33,000 feet, the core recovered silty and clayey sediments with a little sand that included fragments of metamorphic rock like that known from eastern Mindanao. This core, 20 inches long, showed no signs of sliding sediment.

These indications of long suspension and wide dispersal of land-derived sediments confirm the evidence of strong currents at great depth. It appears that the currents cross the ridges and saddles, winnow out the fine material settling there, and waft it along until it falls either into one of the basins or off the continental slope.

On many of the steep slopes bare rock has been dredged at intervals. Nearby the sediments may be coarse and unsorted at depths of thousands of feet. Gentler slopes at the same depths have much finer sediments and no bare rock; it thus appears that some of these coarse sediments may have slid down the steep submarine slopes from higher levels, exposing the rock and coming to rest at depths where the normal sediment is far finer. Submarine slumping like that inferred here is well shown in consolidated rocks by contorted bedding and sliding surfaces, as in the Silurian rocks of parts of Wales, in Tertiary rocks of Peru and Ecuador, and in many other places.

On the divide between two of the basins, the "Snellius" made current readings at depths ranging from the surface to 6,500 feet. Currents ranged in direction from NE to E to SE and even SW at different depths; in speed they ranged from S!3 mile per hour at the surface to 1 mile per hour at 600 feet, whence they fell off to 1/14 mile per hour at 6,500 feet. From experiments it has been found that particles of quartz of various diameters settle in water at speeds indicated in the accompanying table.

Miss Neeb, a Dutch geologist who studied the sediments, computed the distance they would be transported by currents of only 1 centimeter per second (1/44 mile per hour) during the time required for them to settle 1,000 meters (3,300 feet).

In this calculation she assumed no turbulence, which would, of course, further prolong the suspension of some particles, as would also the higher viscosity and density of sea water, at the lower temperatures which prevail at depth.

Table: Theoretical Dispersal of Sedimentary Particles by Currents

Diameter of particle (mm.)	*Rate of settling in water at 27° c. (cm./hr.)*	*Distance transported while sinking 1,000 meters and being carried by current of 1 cm./sec. (km.)*
0.100 (fine sand)	2000	1.8
0.030 (coarse silt)	180	20
0.005 (fine silt)	5	720
0.001 (clay)	0.2	18,000

These computations, though they neglect flocculation and therefore cannot be literally accepted, make it appear probable that the finest silts and clays can be carried almost indefinitely by the currents actually found in the East Indies; in fact, the clay particles in the sediments of a basin may have sources quite different from those of the coarser silts and sand grains with which they are deposited.

Still another result of the studies of the East Indian sediments is noteworthy. The bottom muds near Java are made up of volcanic debris from the nearby island; farther north, the sediments are coarser quartz sands, undoubtedly derived from the deeply weathered granites of Borneo and the islands north of Sumatra.

Because they are separated from these quartz-bearing islands by a belt of much finer and presumably younger quartzose sands, they have been considered as residual from the drowned land surface, relatively undisturbed by marine processes. Although the erratic size-distribution of many of our Atlantic shelf deposits makes this deduction somewhat doubtful, it may be that the sea, during its rise since the Pleistocene glaciers began to melt, has indeed only slightly reworked these sands.

Pelagic Sediments

Pelagic sediments cover nearly three quarters of the ocean bottom, as the "Challenger" discovered. Deep-sea deposits are of three main types: the *calcareous oozes,* the *red clays,* and the *siliceous oozes.*

Of these, the *calcareous oozes* are most widespread, covering nearly 48 per cent of the sea floor, followed by the red clay (38%) and siliceous oozes (14%). As we have noticed these sediments are not found near land, even in extremely deep water, because their characteristic materials are masked by a deluge of terrigenous debris. According to recent compilations, the calcareous oozes (812 samples) have been dredged from depths of 2,300 to nearly 20,000 feet, with the average less than 12,000 feet, the siliceous oozes (only 37 samples) from depths between 3,400 and 26,500 feet (average about 14,000), while the red clay (126 samples) has nowhere been found at depths less than 13,000 feet and extends to as much as 27,000 feet.

CALCAREOUS OOZES

Calcareous oozes cover most of the floor of the Atlantic and Indian oceans and of the equatorial and southern Pacific as far as latitude 50° S.

They consist largely of remains of planktonic organisms having calcium carbonate shells. Shells of *Foraminifera* and plates from calcareous algae are most abundant. Minute particles of clay and subordinate siliceous shells are present. As noted above, the calcareous oozes are, on the average, the shallowest of the pelagic sediments.

They are considered to be forming where, though some terrigenous clays are concurrently settling, the calcareous plankton can rain down on the ocean floor in sufficient quantity to mask the clay. The relatively high content of carbon dioxide below the zone of photosynthesis (where it is largely consumed by plants) makes deep-sea water a slow solvent of the carbonate shells. The ooze, nevertheless, accumulates because the falling material buries the underlying sediments and protects them, for the solution is slow and the interstitial water diffuses so sluggishly that it soon becomes saturated with carbonate.

Red Clay

Red clay covers most of the deep-ocean floor. Most of it is too fine grained to be identified microscopically, but X-ray study has shown that the chief minerals are quartz, mica, and several kinds of clay minerals. All are generally stained red by iron rust. None of the truly oceanic islands has any quartz rocks, and the clay minerals are those found in soils formed on land, hence much of the material must have been derived from the continents.

Volcanic dust may also be an important source. The red clays contain a few siliceous organisms—diatoms or radiolaria—and also a few of those so prominent in calcareous oozes. In fact, all three types of pelagic deposits intergrade, and their distinctions are arbitrary.

It has been noticed that red clay accumulates either where few calcareous planktonic organisms can thrive (in high latitudes) or where such organisms must sink through vast depths to reach bottom. The deep water is cold and carries considerable carbon dioxide in solution. Thus most of the calcareous materials dissolve during sinking, and not enough reach the bottom to mask the clay that is everywhere accumulating.

There is little organic matter, so that the oxygen in the water can combine with iron to give the characteristic red hue. Manganese oxide is abundant in some of the red clay, and its deposition is not fully understood, but it is thought to be somehow related to the oxidizing condition at the bottom.

Siliceous Oozes

Siliceous oozes carpet much of the polar seas and an equatorial strip across most of the eastern Pacific. The equatorial oozes are largely composed of shells of radiolaria, minute single-celled animals, whereas those of high latitudes are chiefly shells of the single-celled plants, the diatoms.

Radiolaria flourish in the nutritious water of the equatorial drift. In high latitudes the mixing of deep water with cold surface water and the excessively deep turbulence caused by the prevailing westerlies furnish abundant food for plankton. Furthermore, the waters are cold and rich in carbon dioxide, so that most limy shells are dissolved before reaching bottom. Diatoms, therefore, accumulate here so abundantly as to mask other constituents of the ooze.

Pelagic Sediments in the Geologic Record

Claystone deposits so similar to pelagic red clays (manganese nodules, very fine grain, rich in iron, with fragmental foraminifera and other features) that some geologists consider them deep-sea deposits, have been described from the Alps, from Barbados in the West Indies, and from Timor in the East Indies. These rocks are so intimately associated with coarse clastics and rocks with other shallowwater characteristics, however, that a deep-sea origin seems doubtful, though not impossible. The Cretaceous chalk of England and France was at one time also considered a consolidated deep-sea ooze. Its fossils are largely shallowwater forms, however, and it is now thought that the chalk was probably deposited in clear water at depths not exceeding a few hundred feet.

The Cretaceous rocks of the Caspian region include a great thickness of coccolith-bearing chalks, associated with narrow zones of terrigenous ("blue") mud shales, and, apparently, shallow benthic fossils are lacking. Whether or not these particular beds do, indeed, represent a fossil ooze, it seems certain that deep-sea sediments are extremely rare on land.

DEPOSITION OF LIMESTONE AND DOLOMITE

Limestones and dolomites make up about 20 per cent of the sedimentary rocks. Today, as we have just noted, calcareous oozes cover nearly half the ocean bottom. These pelagic oozes, however, are only subordinately, if at all, represented on land, and certainly most limestones have been formed in shallow water, as is shown by their fossils. Among modern sediments we find many analogs to these shallow marine limestones but not to the dolomites.

CHEMICAL CONDITIONS AFFECTING CALCITE DEPOSITION

Calcite is among the most soluble of the common minerals. Though only slowly soluble in pure water, it readily dissolves in the presence of carbon dioxide.

The solubility of calcite in natural sea water is difficult to determine because of the interference of so many other dissolved ions. Despite generations of chemical work, recent studies still differ by a factor of 6: according to one, sea water in contact with air, and both at a temperature of 68° F., is about 6-fold oversaturated; according to another, it is about saturated. The latter result seems far more credible, for oversaturation could hardly continue in a sea constantly stirred and with many calcite shells available as nuclei of precipitation.

Despite the complexity of the problem, all chemists agree that the warming of sea water, which consequently decreases its carbon dioxide content and, usually, increases salinity, is favorable to precipitation of calcite. On the other hand, decomposition of organic matter, which yields carbon dioxide, increases the solubility of calcite. The composition of the pelagic sediments suggests that the great bulk of the ocean is slightly undersaturated, but that the warm surface waters of low latitudes are nearly or quite saturated with calcite.

Life processes affect marine deposition and solution of calcite by modifying the carbon dioxide content of the water. Many organisms secrete shells of calcite or aragonite (a different crystalline form of calcium carbonate). Photosynthesis by benthic and pelagic plants extracts carbon dioxide from the water; in the light zone of the warm tropical seas the water is practically saturated.

Over the shallow Bahama Banks many processes unite to reduce the carbon dioxide content: The water is warmed as it flows over the bank to replenish that lost by evaporation; flourishing plants withdraw much more carbon dioxide;

bacteria in the muds include species that turn nitrogen into ammonia. Ammonia, being a weak base, neutralizes some of the free hydrogen ions in the water and thus further reduces the solubility of calcium carbonate. As a result of all these processes, the water is saturated, and fine needles of calcium carbonate are constantly being formed and settling to the bottom as lime muds. It seems likely that some of the extremely fine-grained limestones of the geologic past were formed similarly.

Many lime-secreting plants and animals flourish in the littoral zone of tropical seas. Shells and their broken fragments are so abundant in many places as to make up whole beaches. They also accumulate in shallow offshore waters and are slowly cemented to limestone.

Coral Reefs

Conspicuous among calcareous accumulations are the coral reefs. They are confined to waters at least as warm as 68° F., for reef-building corals do not thrive in colder water. Not only must the water be warm, but it must also have normal salinity and be nearly free from silt; coral reefs are therefore absent near the mouths of muddy rivers. The coral animal, anchored in his limy case, depends upon food brought to him by the waves and currents and therefore thrives better on the windward and offshore sides of the reefs. Many species of corals live on the reefs, but reef-builders will not grow below a depth of about 150 feet nor much above low-tide level. Because of their narrow depth range, coral reefs are delicate markers of crustal movement.

Though corals form the conspicuous framework of the reefs, algal remains bind them together, and many other calcareous organisms such as mollusks, worms, foraminifera, and a host of others contribute to the reef and may constitute far more of its bulk than coral.

Coral reefs are of three kinds: *fringing reefs, barrier reefs,* and atolls. Fringing reefs are confined to the very border of the land. They may be thousands of feet wide, but most are only a few score feet. In the aggregate, however, they are undoubtedly much bulkier than the more spectacular and individually larger barrier reefs and atolls.

The Great Barrier Reef of Australia roughly parallels the Queensland Coast at distances ranging from 25 to nearly 200 miles and extends for 1,200 miles south from Torres Strait. The barrier reef is separated from shore by lagoons which may be shallow but are in places hundreds of feet deep.

Many of the volcanic islands of the Pacific and Indian oceans are surrounded by barriers. In atolls there is no central island, only a ringlike reef enclosing a central lagoon. The enclosed lagoon is dotted with isolated coral heads and floored with algal mud and some sand derived from broken reef fragments hurled over by the waves.

The Origin of Atolls

The origin of barrier reefs and atolls has fascinated students for more than a century, since Darwin's "Voyage of the 'Beagle'." Darwin considered fringing reefs, barrier reefs, and atolls to form a sequence. He thought that if an island fringed by a reef subsided slowly, the reef would grow upward essentially in place, keeping pace with the subsidence to become a barrier reef, and when subsidence had drowned the central island, an atoll. In support of this, it has been pointed out that the shore forms of many islands with fringing reefs are indented like those that might be developed by submergence of a stream-dissected mountain. Furthermore, the outer slopes of many atolls and reefs descend steeply to great depths, and at least two borings on reefs show their structure to be chiefly coralline to depths of many hundred feet.

An alternate hypothesis was suggested by the American geologist Daly in 1910. Daly noticed that charts of the lagoons of many barrier reefs and atolls show a remarkably uniform depth, ranging from about 150 to 250 feet. He also noted that some of the higher peaks of the Hawaiian Islands had been glaciated, presumably during the Pleistocene, and inferred that the water along the shore below must have been too cool for coral growth at that time. If so, the present fringing reefs of

Hawaii must all date from postglacial time. He then computed the volume of sea water that must have been locked up in the ice sheets during their greatest expansion. From this he inferred that sea level was then lowered about 300 feet beneath its present stand. If so, the water in which the corals now flourish must have been not only colder, but also siltier, because the waves would be working on unconsolidated sediments of the sea floor.

Therefore, Daly concluded, the reefs would not have been able to withstand the attack of the sea—they should have been planed off at a level appropriate to the lowered sea of the time. When the ice sheets melted, and sea level gradually rose, the surviving corals would find smoothly planed banks to colonize. According to this theory, then, the atolls and barrier reefs have been built on smoothly leveled surfaces of abrasion as the water gradually rose.

No subsidence of the land is needed to explain the lagoons. The present lagoons vary in depth because of postglacial sedimentation, the smaller lagoons are shallower than the larger because their areas are proportionately less, relative to the reefs over which sediment is washed during storms. Banks and sea mounts that have not been colonized by reef-builders since the rise of sea level are simply the planed-off islands of preglacial time.

Daly's theory fails to explain the indented shores of many islands within barrier reefs, for the relief of the valleys and ridges is far greater than can be explained by a mere 300-foot drop in sea level. On the other hand, it does satisfactorily explain the uniform depths of lagoons. Probably both Darwin and Daly are partly correct in explaining the origin of fringing reefs and atolls, for certainly not all parts of the ocean floor have had the same history, even since the Pleistocene. For instance, we know of both drowned reefs with perfectly preserved atoll shapes near the Philippines and of greatly elevated reefs in Fiji and Samoa.

Dolomites

Dolomites offer one of the greatest of geological puzzles.

Very few organisms secrete shells with appreciable magnesium carbonate content, and in the few that do, the ratio of magnesium carbonate to calcium carbonate is far short of that in dolomite. We know of no deposits of inorganic dolomite on the present sea floor. Yet dolomites, though practically absent among the younger rocks, are both abundant and widespread among Mesozoic and older rocks.

Locally, around certain igneous intrusions, limestone has been altered over considerable areas to dolomite. This dolomite shows, however, definite association with faults and fissures and is clearly a product of contact metamorphism. Such an origin is not possible for the great sheets of early Paleozoic dolomite that can be traced for scores or even hundreds of miles, some of them interbedded in the closest association with unaltered limestones.

A possible clue to the origin of these widespread dolomites comes from a boring on the atoll of Funafuti, in the Ellice Islands, not far from Tarawa of World War II fame. In this boring, which reached a depth of about 1,100 feet, all the material greatly resembled that at the surface—that is, coral heads, algal deposits, and shells of many other animals. However, although the structures of these organic remains were well preserved, the composition of the material at depth differed notably from that at the surface.

The content of magnesium carbonate irregularly increased with depth, and considerable dolomite was present near the bottom. It was inferred that part of the calcium in the calcite shells had been replaced by magnesium ions from the sea water, which, of course, thoroughly soaks the atoll structure. Magnesium is five times as abundant as calcium in sea water. This hypothesis may be correct, but, if so, other factors are locally involved that prevent its occurrence everywhere, for a core hole drilled on Bikini atoll to a depth of 2,200 feet found no dolomite or any significant increase in magnesium content, even though the lower beds penetrated had fossils of Miocene age.

These beds must have been in essentially continuous contact with sea water for more than 15 million years. The details of the origin of dolomite offer some of the many geologic problems for the future.

SUBMARINE CANYONS

Terrigenous sediments are mostly deposited nearshore, as shown by the great thicknesses—measured in thousands to tens of thousands of feet—of Cenozoic and Cretaceous shale and sandstone penetrated by many coastal wells, and by the fineness and thinness of the elastic layers in deep-sea deposits. The continental shelves appear to be primarily constructional terraces, built up through the later part of geologic time; and, secondarily, erosional features cut on rocks of varied kinds and ages.

Submarine erosion may not, however, be limited by wave base. Since 1890 increasing knowledge of the sea floor has revealed more and more signs of deeper erosion, notably, valleys cut sharply into the shelf. Note that the only river mouth clearly connected with a submarine canyon is the Hudson's. This exceptional trench extends from the estuary of the Hudson River almost across the shelf and, then, after fading out on the shelf, reappears as a deep trench in the continental slope, extending clear to the floor of the Atlantic.Most submarine canyons are short and steep features of the continental slope, heading far from shore except where the shelf is very narrow or interrupted by deep basins, as off southern California. A few canyons are 70 to 150 miles long and 3,000 to 5,000 feet deeper than the adjacent sea floor, thus comparable to the Grand Canyon in size. Among the great submarine canyons are the Monterey Canyon off California that heads near the mouth of Salinas River, the Congo Canyon extending into the Atlantic from the mouth of the Congo River, and the Nazaré Canyon off Portugal that heads far from any existing river mouth.

Monterey Canyon's long profile, together with that of the Salinas River. The irregularities in the submarine profile may be caused by errors in the soundings, but the general form is trustworthy. The average grade of the canyon bottom is almost 4° for the first 40 miles from the shore, that of the lower Salinas

River less than 0.1°. The submarine canyon is probably cut almost entirely in Miocene, Pliocene, and Pleistocene sediments, though near mid-course granite has been found on one wall. The tributary Carmel Canyon may be excavated chiefly in sheared granite. The head of Monterey Canyon is just outside the beach zone, apparently cut into latest Pleistocene or even recent marine and river deposits.

In general, submarine canyons are marked by steep gradients, roughly concave long profiles, few tributaries, and they head within a few hundred feet of sea level. The relatively small number of canyons whose heads are near the mouths of land rivers include most of the very large examples. The canyon mouths have not been adequately surveyed. Some are at the levels of local basins at various depths below 1,200 feet, or at the base of the steep continental slope 7,000 to 11,000 feet subsea, or, in the case of Hudson Canyon on the deep-sea floor, at even greater depth. The walls are composed chiefly of the younger sedimentary rocks, in part, fairly well consolidated.

Hypothesis of Subaerial Origin

The many hypotheses advanced to account for submarine canyons belong to two groups, the subaerial and the submarine. The principal subaerial hypothesis makes use of the sea level lowering during the Pleistocene glacial epochs. The question here is the amount of water removed from the oceans and locked up in glacial ice. Although the areal extent of the Pleistocene glaciers is fairly well known, their thickness is uncertain, so estimates of the ice's volume vary.

Perhaps the lowering of sea level can be determined more easily from direct topographic evidence. As sea level is both the locus of wave erosion and the base level for stream erosion, there would be two sets of topographic effects. After a lowering of sea level, wave erosion would have had new and weak materials on which to work. Wave-cut terraces would therefore be ex pected on the continental slope. All trunk rivers would have been rejuvenated. The Mississippi, Colorado, Yangtze, Ganges, Indus, and La Plata would have been among the great rivers especially effective in cutting valleys into the newly exposed weak sediments of their lower courses. Even the Salinas River could have cleaned out its lower valley with ease for the first 3,000 feet of depth, and with little difficulty down to 8,000 feet subsea. The valleys so excavated might have been filled with continental or marine sediments as base level rose again, but the deep fill would then tell the tale. The long profile of the part of the canyon still unfilled should pass beneath the base of the fill in a smooth curve. This smooth profile curve might extend inland 1,000 miles for the Mississippi and perhaps 50 miles for the Salinas River.

What are the actual topographic facts? Terraces submerged several thousand feet may have been detected off Angola, west central Africa, but they have not been found elsewhere. Two isolated sea mounts 600 and 800 miles off the California coast have flat tops 1,700 and 2,400 feet below sea level. There is no known example of a smooth long profile with a 6,000-foot, 7,000foot, or other similar deep base level, extending inland along an estuary or along the base of Pleistocene alluvial or marine fill, though the Congo may be a possible imperfect example.

On the other hand, all the topographic features that would be caused by submergence have been found, related to a base level only a few hundred feet below present sea level. The floors of atolls 200 or 300 feet subsea, and submerged drainage systems, deltas, and even beaches at about the same depths. Valleys floored with Pleistocene or Recent alluvium 200 to 600 feet deep are known along many coasts and may be generally distributed. After allowances for local deformations and for the depths of great rivers, the evidence seems to indicate that the lowest Pleistocene sea levels were about 300 feet below the present one, in good accord with reasonable estimates of the volume of the ice.

Another effect of a theoretical drop of sea level by 6,000 or 7,000 feet deserves mention. The volume of the ocean would be reduced about one-half, with a corresponding doubling of the salinity. The more sensitive forms of marine life, including many shelled animals, would be destroyed. However, only a few kinds of shellfish became extinct during the Pleistocene, and the number of these extinctions appears to be merely normal, as it corresponds nicely with estimates of the period's length based on other grounds. Apparently, not one of the corollaries of this subaerial hypothesis is actually true.

HYPOTHESIS OF SUBMARINE FORMATION BY DENSITY CURRENTS

Turbid water is denser than clear water. A turbid water mass moves along a sloping bottom as a density current. These currents have been studied experimentally, especially by Kuenen of Holland. The motive power of such a current is gravitational, of course, and is measured by the difference between its density and that of the main water body, and by the slope. Since the difference in density is usually much less than that between water and air, a greater slope is required to produce a given velocity. Even boulders, however, if suspended in a muddy fluid, move rapidly down 3° or 4° slopes.

The best historical evidence for erosion by currents comes from Swiss lakes. The mean gradient of 4° is similar to that of many submarine canyons, and is much higher than most stream gradients on land. The largest trench in a Swiss lake is that cutting the delta of the upper Rhone in Lake Geneva. American reservoirs, such as Lake Mead on the Colorado River, also have their density currents, which travel tens of miles along the lake floor.

In a fresh-water lake a density current sinks to the bottom if its density exceeds 1. In the sea the density must be greater than 1.025, the value for clear sea water. Bottom density currents in the sea are obviously hard to discover. The evidence for their effectiveness in cutting canyons must be mostly circumstantial. We would expect steep gradients; canyon heads just below sea level; mouths at various levels, perhaps especially at the foot of the continental slope; and a distribution of canyons determined by the sources and movements of abundant supplies of sediments.

A large muddy river would be one good source of sediment, the meeting place between two longshore currents another, the outer edge of the continental shelf a third-especially good during the Pleistocene stages of lowered sea level and abundant debris from melting glaciers. Once started, erosion of the canyons themselves would furnish additional sediment.

If density currents cut deep submarine canyons, they must also be responsible for some marine deposits, and ancient deposits of this sort might, after elevation and erosion, be exposed to our view. Just such contorted and tumbled, medium-to-coarse fragmental deposits have been found in Italian and Californian Cenozoic strata and are less certainly represented in many other formations of all eras, including the pre-Cambrian.

Nearly all the known facts about submarine canyons seem to fit the density current hypothesis. Nevertheless, two or three of the best-informed American students of the subject still favour a modified subaerial hypothesis. One modification involves deformation after canyon formation. The canyons are thought to have been bowed down as the continents spread laterally under their own weight. The resulting long profiles would be convex. We cannot feel really satisfied with any hypothesis yet; more information is needed.

Can hard rocks be eroded by density currents? What is the form of the deep-sea floor? What is the significance of the flat-topped sea mounts? Are there terraces at or near the base of the continental slope? How thick is the sedimentary layer on the deep-sea floor? Do debris deltas extend out from the mouths of submarine canyons? Some of these questions have been answered partially and locally, but many more sonic and seismic surveys are needed, as well as more long cores from the uppermost deep-sea sediments.

Perhaps trough experiments, or instruments planted in the canyons themselves, will measure density-current erosion. Ultimately, the solution of this one problem will give us firm ground for further advances in our knowledge of both oceans and rocks.

Chapter 5

Theory of Soil Science

Soil,' the term generally applied to that part of the earth's ' This word comes through O. Fr. *soil* from a Late Latin usage of *solea* for soil or ground, which in classic Lat. meant the sole of the foot, also a sandal. This was due to a confusion with *solum,* ground, whence Fr. *sol.* Both *solea* and *solum* are, of course, from the same root. To be distinguished from this word is " soil," to make dirty, to stain, defile. The origin is the O. Fr. *soil* or *souil,* the miry wallowing ground of a wild boar, whence the hunting phrase " to take soil," of a beast of the chase taking to water or marshy ground.

The derivation is therefore from Lat. *soillus,* pertaining to substance which is stirred or tilled by implements such as ploughs and spades. Below this is the *subsoil.* The soil through being acted upon by the air, heat, frost and other agencies usually consists of finer particles than those. comprising the bulk of the subsoil. It contains more roots, and as a rule, is darker in colour than the subsoil on account of the larger proportion of decaying vegetable matter present in it: it is also looser in texture than the subsoil.

The subsoil not unfrequently contains materials which are deleterious to the growth of crops, and roots descending into it may absorb and convey these poisonous substances to other parts of the plant or be themselves damaged by contact with them. On this account deeper tillage than usual, which allows of easier penetration of roots, or the carrying out of operations which bring the subsoil to the surface, must always be carefully considered. At first sight few natural materials appear to be of less interest than the soil; yet its importance is manifest on the slightest reflection. From it, directly or indirectly, are obtained all food materials needed by man and beast. The inorganic materials within it supply some of the chief substances utilized by plants for their development and growth, and from plants animals obtain much of their sustenance.

ORIGIN OF THE SOIL

It is a matter of common observation that stones of monuments, walls or buildings which are exposed to the air sooner or later become eaten away or broken up into small fragments under the influence of the weather. This disintegration is brought about chiefly by changes in temperature, and by the action of the rain, the oxygen, and the carbon dioxide of the air.

During the daytime the surface of the stone may become very warm, while at night it is speedily cooled. Such alterations in temperature produce strains which frequently result in the chipping off of small fragments of the material composing the stone. Moreover the rain penetrates into the small interstices between its particles and dissolves out some of the materials which bind the whole into a solid stone, the surface then becoming a loose powdery mass which falls to the ground below or is carried away by the wind.

The action of frost is also very destructive to many stones, since the water within their cracks and crannies expands on freezing and splits off small pieces from their surfaces. In the case of limestones the carbon dioxide of the air in association with rain and dew eats into them and leads to their disintegration. The oxygen of the air may also bring about chemical changes which result in the production of soluble substances removable by rain, the insoluble parts being left in a loosened state.

These " weathering " agents not only act upon stones of buildings, but upon rocks of all kinds, reducing them sooner or later into a more or less fine powder. The work has been going on for ages, and the finely comminuted particles of rocks form the main bulk of the soil which covers much of the earth's surface, the rest of the soil being composed chiefly of the remains of roots and other parts of plants.

If the whole of the soil in the British Islands were swept into the sea and the rocks beneath it laid bare the surface of the country would ultimately become covered again with soil produced from the rocks by the weathering processes just described. Moreover where there was no transport or solution of the soil thus produced it would necessarily show some similarity in composition to the rock on which it rested. The soils overlying red sandstone rocks would be reddish and of a sandy nature, while those overlying chalk would be whitish and contain considerable amounts of lime. In many parts of the country soils exhibiting such relationships, and known as *sedentary* soils, are prevalent, the transition from the soil to the rock beneath being plainly visible in sections exposed to view in railway cuttings, quarries and other excavations. The upper layer or soil proper consists of material which has been subjected swine, *sus.* " To sully," to besmirch, to cover with " mire " to ages of weathering; the bulk of it is composed of finely comminuted particles of sand, clay and other minerals, among which are imbedded larger or smaller stones of more refractory nature. On descending into the substratum the finer material decreases and more stones are met with; farther down are seen larger fragments of unaltered rock closely packed, and this brash or rubble grades insensibly into the unbroken rock below.

In many districts the soil is manifestly unconnected in origin with the rock on which it rests, and differs from it in colour, composition and other characters. There are *transported* or drift soils, the particles of which have been brought from other areas and deposited over the rocks below. Some of the stiff boulder clays or " till " so prevalent over parts of the north of England appear to have been deposited from ice sheets during the glacial period. Perhaps the majority of drift soils, however, have been moved to their present position by the action of the water of rivers or the sea.

As fast as the rock of a cliff is weathered its fragments are washed to the ground by the rain, and carried down the slopes by small streams, ultimately finding their way into a river along which they are carried until the force of the water is insufficient to keep them in suspension, when they become deposited in the river bed or along its banks. Such river-transported material or *alluvium* is common in all river valleys. It is often of very mixed origin, being derived from the detritus of many kinds of rocks, and usually forms soil of a fertile character.

QUALITY OF SOIL

The good or bad qualities of a soil have reference to the needs of the crops which are to be grown upon it, and it is only after a consideration of the requirements of plants that a clear conception can be formed of what characters the soil must possess for it to be a suitable medium on which healthy crops can be raised.

In the first place, soil, to be of any use, must be sufficiently loose and porous to allow the roots of plants to grow and extend freely. It may be so compact that root development is checked or stopped altogether, in which case the plant suffers. On the other hand it should not be too open in texture or the roots do not get a proper hold of the ground and are easily disturbed by wind: moreover such soils are liable to blow away, leaving the underground parts exposed to the air and drought.

The roots like all other parts of plants contain protoplasm or living material, which cannot carry on its functions unless it is supplied with an adequate amount of oxygen: hence the necessity for the continuous circulation of fresh air through the soil. If the latter is too compact or has its interstices filled with carbon dioxide gas or with water - as is the case when the ground is water-logged - the roots rapidly die of suffocation just as would an animal under the same conditions. There is another point which requires attention. Plants need very considerable amounts of water for their nutrition and growth; the waterholding capacity is, therefore, important. If the soil holds too much it becomes water-logged and its temperature falls below the point for healthy growth, at any rate of the kinds of plants. usually cultivated on farms and in gardens. If it allows of too free drainage drought sets in and the plants, not getting enough water for their needs, become stunted in size. Too much water is bad, and too little is equally injurious.

In addition, the temperature of the soil largely controls the yield of crops which can be obtained from the land. Soil whose temperature remains low, whether from its northerly aspect or from its high water content or other cause, is unsatisfactory, because the germination of seeds and the general life processes of plants cannot go on satisfactorily except at certain temperatures well above freezing-point.

A good soil should be deep to allow of extensive root development and, in the case of arable soils, easy to work with implements. Even when all the conditions above mentioned in regard to texture, water-holding capacity, aeration and temperature are suitably fulfilled the soil may still be barren: plant foodmaterial is needed. This is usually present in abundance although it may not be available to the plant under certain circumstances, or may need to be replenished or increased by additions to the soil of manures or fertilizers.

CHIEF CONSTITUENTS OF THE SOIL

An examination of the soil shows it to be composed of a vast number of small particles of sand, clay, chalk and humus, in which are generally imbedded larger or smaller stones. It will be useful to consider the nature of the four chief constituents just mentioned and their bearing upon the texture, water-holding capacity and other characters which were referred. Sand consists of grains of quartz or flint, the individual particles of which are large enough to be seen with the unaided eye or readily felt as gritty grains when rubbed between the finger and thumb. When a little soil is shaken up with water in a tumbler the sand particles rapidly fall to the bottom and form a layer which resembles ordinary sand of the seashore or river banks. Chemically pure sand is silicon dioxide (SiO_2) or quartz, a clear transparent glass-like mineral, but as ordinarily met with, it is more or less impure and generally coloured reddish or yellowish by oxide of iron. A soil consisting of sand entirely would be very loose, would have little capacity to retain water, would be liable to become very hot in the daytime and cool at night and would be quite unsuitable for growth of plants.

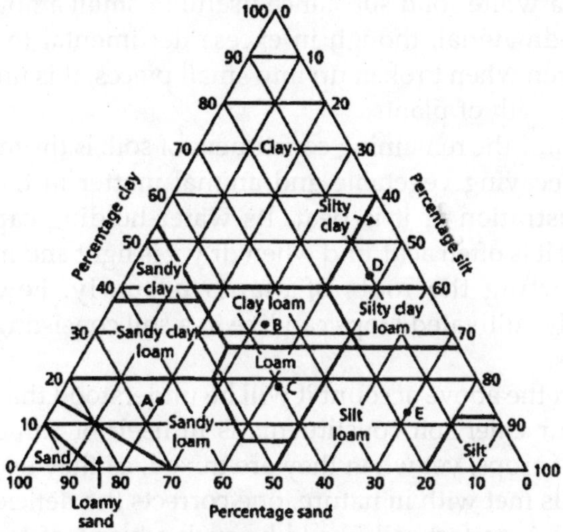

Fig. Constituents of the Soil

The term clay is often used by chemists to denote hydrated silicate of alumina (Al_2O_3 $2SiO_2.2H_2O$), of which kaolin or china clay is a fairly pure form. This substance is present in practically all soils but in comparatively small amounts. Even in the soils which farmers speak of as stiff clays it is rarely present to the extent of more than I or 2%. The word " clay " used in the agricultural sense denotes a sticky intractable material which is found to consist of exceedingly fine particles (generally less than. 005 mm. in diameter) of sand and other minerals derived from the decomposition of rocks, with a small amount of silicate of alumina.

The peculiar character which clay possesses is probably due not to its chemical composition but to its physical state. When wet it becomes sticky and almost impossible to move or work with farm implements; neither air nor water can penetrate freely. In a dry state it becomes hard and bakes to a brick. It holds water well and is consequently cold, needing the application of much heat to raise its temperature. It is obvious, therefore, that soil composed entirely of clay is as useless as pure sand so far as the growth of crops upon it is concerned.

Chalk consists, when quite pure, of calcium carbonate ($CaC0_3$), a white solid substance useful in small amounts as a plant foodmaterial, though in excess detrimental to growth. Alone, even when broken up into small pieces, it is unsuitable for the growth of plants.

Humus, the remaining constituent of soil, is the term used for the decaying vegetable and animal matter in the soil. A good illustration of it is peat. Its water-holding capacity is great, but it is often acid, and when dr y it is light and incapable of supporting the roots of plants properly. Few of the commonly cultivated crops can live in a soil consisting mainly of humus.

From the above account it will be understood that not one of the four chief soil constituents is in itself of value for the growth of crops, yet when they are mixed, as they usually are in the soils met with in nature, one corrects the deficiencies of the other. A perfect soil would be such a blend of sand, clay, chalk and humus as would contain sufficient clay and humus to prevent drought, enough sand to render it pervious to fresh air and prevent waterlogging, chalk enough to correct the tendency to acidity of the humus present, and would have within it various substances which would serve as food-materials to the crops.

Generally speaking, soils containing from 30 to 50% of clay and 50 to 60% of sand with an adequate amount of vegetable residues prove the most useful for ordinary farm and garden crops; such blends are known as " loams," those in which clay predominates being termed clay loalns, and those in which the sand predominates sandy roams. " Stiff clays " contain over 50% of clay; " light sands " have less than to %. In the mechanical analysis of the soil, after separation of the stones and fine gravel by means of sieves, the remainder of the finer earth is subjected to various processes of sifting and deposition from water with a view of determining the relative proportions of sand, silt and clay present in it. Most of the material termed " sand " in such analyses consists of particles ranging in diameter from.5 to. 05 mm., and the " silt " from 05 to 005 mm., the " clay " being composed of particles less than.005 mm. in diameter.

The proportional amount of these materials in a sandy soil on the Bagshot beds and a stiff Oxford clay is given below: - The pore-space within the soil, *i.e.* the space between the particles composing the soil, varies with the size of these particles and with the way they are arranged or packed. It is important, since upon it largely depends the movement of air and water in the land. It is generally from 30 to 50% of the total volume occupied by the soil.

Where the soil grains are quite free from each other the smaller grains tend to fill up the spaces between the larger ones; hence it might be concluded that in clays the amount of pore-space would be less than in coarser sands. This is the case in " puddled " clays, but in ordinary clay soils the excessively minute particles of which they largely consist tend to form groups of comparatively large composite grains and it is in such natural soils that the pore-space is largest.

CHEMICAL COMPOSITION OF THE SOIL

It has been found by experiment that plants need for their nutritive process and their growth, certain chemical elements, namely, carbon, hydrogen, oxygen, nitrogen, sulphur, phosphorus, potassium, magnesium, calcium and iron. With the exception of the carbon and a small proportion of the oxygen and nitrogen, which may be partially derived from the air, these elements are taken from the soil by crops. The following table shows the amounts of the chief constituents removed by certain crops in lb per acre: - Plants also remove from the soil silicon, sodium, chlorine, and other elements which are, nevertheless, found to be unessential for the growth and may therefore be neglected here.

Leguminous crops take some of the nitrogen which they require from the air, but most plants obtain it from the nitrates present in the soil. The sulphur exists in the soil chiefly in the form of sulphates of magnesium, calcium and other metals; the phosphorus mainly as phosphates of calcium, magnesium and iron; the potash, soda and other bases as silicates and nitrates; calcium and magnesium carbonates are also common constituents of many soils. In the ordinary chemical analyses of the soil determinations are made of the nitrogen and various carbonates present as well as of the amount of phosphoric acid, potash, soda, magnesia and other components soluble in strong hydrochloric acid.

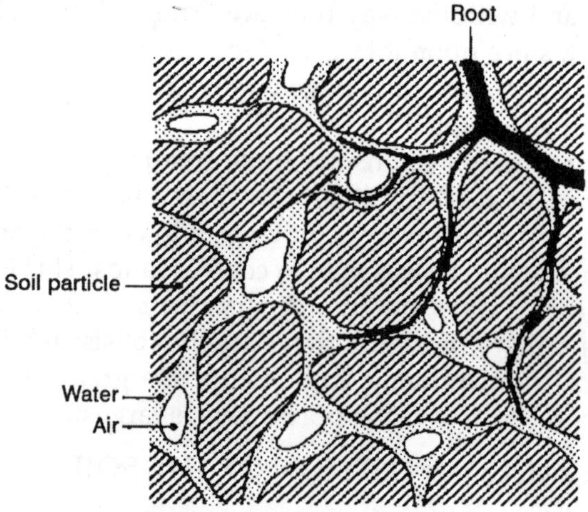

Fig. Composition of the Soil

Below are given examples of the analyses of a poor sandy soil and an ordinary loam: - Since the dry weight of the first foot of soil over an acre is about 4,000,000 lb the poor sandy soil contains within it: Nitrogen. 7,600 lb Phosphoric acid. 7,200 „ Potash.. 7,600 „ Lime 9,200 „ From the figures given previously of the amount of nitrogen, potash and phosphoric acid removed by a wheat or mangel crop it would appear that this soil has enough of these ingredients in it to yield many such crops; yet experience has shown that these crops cannot be grown on such a poor sandy soil unless manures containing phosphates, potash and nitrogen are added.

Many attempts have been made to correlate the results of the analyses of a soil with its known cropping power, but there is yet much to be learnt in regard to these matters. A great proportion of the food constituents which can be extracted by strong hydrochloric acid are not in a condition to be taken up by the roots of plants; they are present, but in a " dormant " state, although by tillage and weathering processes they may in time become " available " to plants. Analyses of this character would appear to indicate the permanent productive capacity of the soil rather than its immediate power of growing a crop.

Soils containing less than 25% of potash are likely to need special application of potash fertilizers to give good results, while those containing as much as. 4 or. 5% do not usually respond to those manures.

Where the amount of phosphoric acid is less than. 05% phosphatic manures are generally found to be beneficial; with more than. 1% present these fertilizers are not usually called for except perhaps in soils containing a high percentage of iron compounds. Similarly soils with less than i % of nitrogen are likely to be benefited by applications of nitrogenous manures. Too much stress, however, cannot be laid upon these figures, since the fertility of a soil is very greatly influenced by texture and physical constitution, perhaps more so by these factors than by chemical composition.

At present it is not possible to determine with accuracy the amount of immediately available plant food-constituents in a soil: no doubt the various species of plants differ somewhat in their power of absorbing these even from the same soil. The method introduced by Dyer of dissolving out the mineral constituents of the soil with a i % solution of citric acid, which represents about the average acidity of the roots of most common plants, yields better results.

In the case of arable soils, where the amount of phosphoric acid determined by this method falls below 01%, phosphatic manuring *is* essential for good crops. The writer has found that many pasture soils containing less than. 025 or 03%, respond freely to applications of phosphates; probably in such cases even the weak acid is capable of dissolving out phosphates from the humus or other compounds which yield little or none to the roots of grasses and clovers. In soils where the potash available to citric acid is less than. 005%, kainit and other potash fertilizers are needed.

WATER IN THE SOIL

The importance of an adequate supply of water to growing crops cannot well be over-estimated. During the life of a plant there is a continuous stream of water passing through it which enters by the root-hairs in the soil and after passing along the stem is given off from the 'stomata of the leaves into the open air above ground. It has been estimated that an acre of cabbage will absorb from the land and transpire from its leaves more than ten tons of water per day when the weather is fine. In addition to its usefulness in maintaining a turgid state of the young cells without which growth cannot proceed, water is itself a plant food-material and as absorbed from the soil contains dissolved in it all the mineral food constituents needed by plants for healthy nutrition.

Without a sufficient supply plants remain stunted and the crop yield is seriously reduced, as we see in dry seasons when the rainfall is much below the average. If one condition is more necessary than another for good crops it is a suitable supply of water, for no amount of manuring or other treatment of the soil will make up for a deficient rainfall.

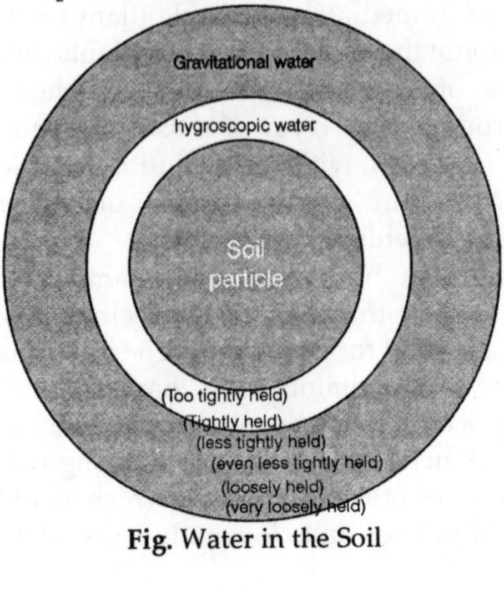

Fig. Water in the Soil

The amount needed for the most satisfactory nutrition varies with different plants. In the case of fair average farm crops it has been shown that for the production of one ton of dry matter contained in them from 300 to 500 tons of water has been absorbed and utilized by the plants. This may be more than the rainfall, in which case irrigation or special control of the water supply may be necessary.

The water-holding capacity of a soil depends upon the amount of free space between the particles of which it is composed into which water can enter. In most cases this amounts to from 30 to 50% of the volume of the soil.

When the pore-space of the soil is filled with water it becomes water-logged and few plants can effect absorption by their roots under such conditions. The root-hairs die from want of air, and the whole plant soon suffers. Fields of wheat and other cereals rarely recover after a week's submergence, but orchards and many trees when at rest in winter withstand a flooded or water-logged condition of the soil for two or three weeks without damage.

The most satisfactory growth is maintained when the amount of water present is not more than 40 to 60% of what would saturate it. Under such conditions each particle of soil is surrounded by a thin film of water and in the pore-space air can freely circulate. It is from such films that the root-hairs absorb all that plants require for their growth. The movement of water into the root-hairs is brought about by the osmotic action of certain salts in their cell-sap. Crops are, however, unable to absorb all the water present in the soil, for when the films become very thin they are held more firmly or cling with more force to the soil particles and resist the osmotic action of the root-hairs. Plants have been found to wither and die in sandy soils containing i a % of water, and in clay soils in which there was still present 8% of water.

When a long glass tube open at both ends is filled with soil and one end is dipped in a shallow basin of water, the water is found to move upwards through the soil column just as oil will rise in an ordinary lamp wick. By this capillary action water may be transferred to the upper layers of the soil from a depth of several feet below the surface. In this manner plants whose roots descend but a little way in the ground are enabled to draw on deep supplies. Not only does water move upwards, but it is transferred by capillarity in all directions through the soil. The amount and speed of movement of water by this means, and the distance to which it may be carried, depend largely upon the fineness of the particles composing the soil and the spaces left between each.

The ascent of water is most rapid through coarse sands, but the height to which it will rise is comparatively small. In clays whose particles are exceedingly minute the water travels very slowly but may ultimately reach a height of many feet above the level of the " water-table " below. While this capillary movement of water is of great importance in supplying the needs of plants it has its disadvantages, since water may be transferred to the surface of the soil, where it evapo rates into the air and is lost to the land or the crop growing upon it.. The loss in this manner was found to be in one instance over a pound. of water per day per square foot of surface, the " water-table " being about 4 or 5 ft. below.

One of the most effective means of conserving soil moisture is by " mulching," *i.e.* by covering the surface of the soil with some loosely compacted material such as straw, leaf-refuse or stablemanure. The space between the parts of such substances is too large to admit of capillary action; hence the water conveyed to the surface of the soil is prevented from passing upwards any further except by slow evaporation through the mulching layer. A loose layer of earth spread over the surface of the soil acts in the same way, and a similarly effective mulch may be prepared by hoeing the soil, or stirring it to a depth of one or two inches with harrows or other implements. The hoe and harrow are therefore excellent tools for use in dry weather. Rolling the land is beneficial to young crops. in dry weather, since it promotes capillary action by reducing the soil spaces. It should, however, be followed by a light hoeing or harrowing.

In the semi-arid regions of the United States, Argentina and other countries where the average annual rainfall lies between ioa to 20 in., irrigation is necessary to obtain full crops every year. Good crops, however, can often be grown in such areas without irrigation if attention is paid to the proper circulation of water in the soil and means for retaining it or preventing excessive loss by evaporation. Of course care must be exercised in the selection of plants - such as sorghum, maize, wheat, and alfalfa or lucerne - which are adapted to dry conditions and a warm climate.

So far as the water-supply is concerned - and this is what ultimately determines the yield of crops - the rain which falls upon the soil should be made to enter it and percolate rapidly through its interstices. A deep porous bed in the upper layers is essential, and this should consist of fine particles which lie close to each other without any tendency to stick together and " puddle " after heavy showers. Every effort should be made to prepare a good mealy tilth by suitable ploughing, harrowing and consolidation.

In the operation of ploughing the furrow slice is separated from the soil below, and although in humid soils this layer may be left to settle by degrees, in semi-arid regions this loosened layer becomes. dry if left alone even for a few hours and valuable water evaporates. into the air. To prevent this various implements, such as disk harrows and specially constructed rollers, may be used to consolidate the upper stirred portion of the soil and place it in close capillary relationship with the lower unmoved layer.

If the soil is allowed to become dry and pulverized, rain is likely to run off or " puddle the surface without penetrating it more than a very short distance. Constant hoeing or harrowing to maintain a natural soil mulch layer of 2 or 3 in. deep greatly conserves the soil water below. In certain districts where the rainfall is low a crop can only be obtained once every alternate year, the intervening season being devoted to tillage with a view of getting the rain into the soil and retaining it there for the crop in the following year.

Bacteria in the Soil

Recent science has made much progress in the

investigation of the micro-organisms of the soil. Whereas the soil used to be looked upon solely as a dead, inert material containing certain chemical substances which serve as food constituents. of the crops grown upon it, it is now known to be a place of habitation for myriads of minute living organisms upon whose activity much of its fertility depends. They are responsible for many important chemical processes which make the soil constituents more available and better adapted to the nutrition of crops. One cubic centimetre of soil taken within a foot or so from the surface contains from II to 2 millions of bacteria of many different kinds, as well as large numbers of fungi. In the lower depths of the soil the numbers. decrease, few being met with at a depth of 5 or 6 ft.

The efficiency of many substances, such as farm-yard manure,. guanos, bone-meal and all other organic materials, which are spread over or dug or ploughed into the land for the benefit of farm and' garden crops, is bound up with the action of these minute living beings. Without their aid most manures would be useless for plant growth. Farm-yard manure, guanos and other fertilizers. undergo decomposition in the soil and become broken down into compounds of simple chemical composition better suited for absorption by the roots of crops, the changes involved being directly due to the activity of bacteria and fungi. Much of the work carried on by these organisms is not clearly understood; there are, however, certain processes which have been extensively investigated and to these it is necessary to refer.

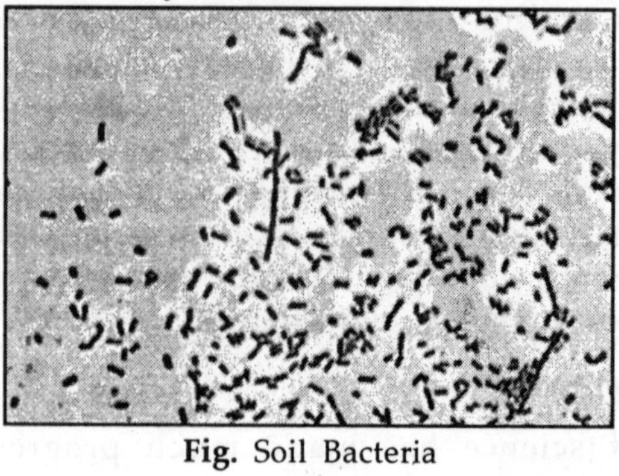

Fig. Soil Bacteria

It has been found by experiment that the nitrogen needed by practically all farm crops except leguminous ones is best supplied in the form of a nitrate; the rapid effect of nitrate of soda when used' as a top dressing to wheat or other plants is well known to farmers.

It has long been known that when organic materials such as the dung and urine of animals, or even the bodies of animals and plants, are applied to the soil, the nitrogen within them becomes oxidized, and ultimately appears in the form of nitrate of lime, potash or some other base. The nitrogen in decaying roots, in the dead stems. and leaves of plants, and in humus generally is sooner or later changed into a nitrate, the change being effected by bacteria. That: the action of living organisms is the cause of the production of nitrates is supported by the fact that the change does not occur when the soil is heated nor when it is treated with disinfectants which destroy or check the growth and life of bacteria.

The process resulting in the formation of nitrates in the soil is spoken of as *nitrification.* The steps in the breaking down of the highly complex nitrogenous proteid compounds contained in the humus of the soil, or applied to the latter by the farmer in the form of dung and organic refuse generally, are many and varied; most frequently the insoluble proteids are changed by various kinds of putrefactive bacteria into soluble proteids, these into simpler amido-bodies, and these again sooner or later into compounds of ammonia. The urea in urine is also rapidly converted by the uro-bacteria into ammonium carbonate.

The compounds of ammonia thus formed from the complex substances by many varied kinds of micro-organisms are ultimately oxidized into nitrates. The change takes place in two stages and is effected by two special groups of nitrifying bacteria, which are present in all soils. In the first stage the ammonium compounds are oxidized to nitrites by the agency of very minute motile bacteria belonging to the genus

Nitrosomonas

The further oxidation of the nitrite to a nitrate is effected by bacteria belonging to the genus *Nitrobacter*. Several conditions must be fulfilled before nitrification can occur. In the first place an adequate temperature is essential; at 5° or 6° C. (40-43° F.) the process is stopped, so that it does not go on in winter. In summer, when the temperature is about 24° C. (75° F.), nitrification proceeds at a rapid rate. The organisms do not carry on their work in soils deficient in air; hence the process is checked in water-logged soils. The presence of a base such as lime or magnesia (or their carbonates) is also essential, as well as an adequate degree of moisture: in dry soils nitrification ceases.

It is the business of the farmer and gardener to promote the activity of these organisms by good tillage, careful drainage and occasional application of lime to soils which are deficient in this substance. It is only when these conditions are attended to that decay and nitrification of dung, guano, fish-meal, sulphate of ammonia and other manures take place, and the constituents which they contain become available to the crops for whose benefit they have been applied to the land.

Nitrates are very soluble in water and are therefore liable to be washed out of the soil by heavy rain. They are, however, very readily absorbed by growing plants, so that in summer, when nitrification is most active, the nitrates produced are usually made use of by crops before loss by drainage takes place. In winter, however, and in fallows loss takes place in the subsoil water.

There is also another possible source of loss of nitrates through the activity of denitrifying bacteria. These organisms reduce nitrates to nitrites and finally to ammonia and gaseous free nitrogen which escapes into the atmosphere. Many bacteria are known which are capable of denitrification, some of them being abundant in fresh dung and upon old straw. They can, however, only carry on their work extensively under anaerobic conditions, as in waterlogged soils or in those which are badly tilled, so that there is but little loss of nitrates through their agency.

An important group of soil organisms are now known which have the power of using the free nitrogen of the atmosphere for the formation of the complex nitrogenous compounds of which their bodies are largely composed. By their continued action the soil becomes enriched with nitrogenous material which eventually through the nitrification process becomes available to ordinary green crops. This power of " fixing nitrogen," as it is termed, is apparently not possessed by higher green plants.

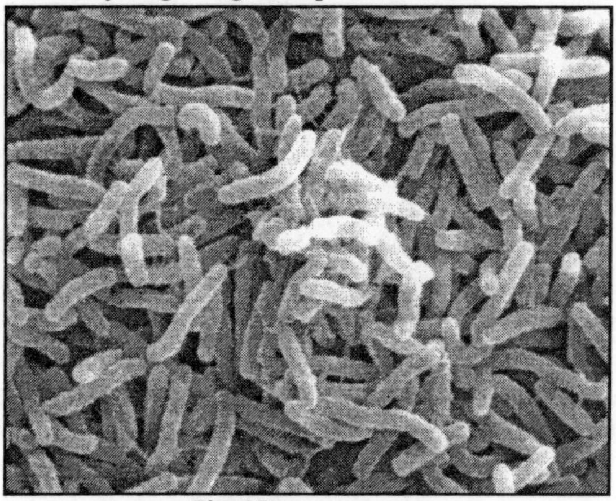

Fig. Nitrosomonas

The bacterium, *Clostridium pasteurianum,* common in most soils, is able to utilize free nitrogen under anaerobic conditions, and an organism known as *Azotobacter chroococcum* and some others closely allied to it, have similar powers which they can exercise under aerobic conditions. For the carrying on of their functions they all need to be supplied with carbohydrates or other carbon compounds which they obtain ordinarily from humus and plant residues in the soil, or possibly in some instances from carbohydrates manufactured by minute green algae with which they live in close union.

Certain bacteria of the nitrogenfixing class enter into association with the roots of green plants, the best-known examples being those which are met with in the nodules upon the roots of clover, peas, beans, sainfoin and other plants belonging to the leguminous order.

That the fertility of land used for the growth of wheat is improved by growing upon it a crop of beans or clover has been long recognized by farmers. The knowledge of the cause, however, is due to modern investigations. When wheat, barley, turnips and similar plants are grown, the soil upon which they are cultivated becomes depleted of its nitrogen; yet after a crop of clover or other leguminous plants the soil is found to be richer in nitrogen than it was before the crop was grown. This is due to the nitrogenous root residues left in the land. Upon the roots of leguminous plants characteristic swollen nodules or tubercles are present. These are found to contain large numbers of a bacterium termed *Bacillus radicicola* or

Pseudomonas Radicicola

The bacteria, which are present in almost all soils, enter the root-hairs of their host plants and ultimately stimulate the production of an excrescent nodule, in which they live. For a time after entry they multiply, obtaining the nitrogen necessary for their nutrition and growth from the free nitrogen of the air, the carbohydrate required being supplied by the pea or clover plant in whose tissues they make a home. The nodules increase in size, and analysis shows that they are exceedingly rich in nitrogen up to the time of flowering of the host plant. During this period the bacteria multiply and most of them assume a peculiar thickened or branched form, in which state they are spoken of as bacteroids.

Later the nitrogen-content of the nodule decreases, most of the organisms, which are largely composed of proteid material, becoming digested and transformed into soluble nitrogenous compounds which are conducted to the developing roots and seeds. After the decay of the roots some of the unchanged bacteria are left in the soil, where they remain ready to infect a new leguminous crop.

The nitrogen-fixing nodule bacteria can be cultivated on artificial media, and many attempts have been made to utilize them for practical purposes. Pure cultures may be made and after dilution in water or other liquid can be mixed with soil to be ultimately spread over the land which is to be infected. The method of using them most frequently adopted consists in applying them to the seeds of leguminous plants before sowing, the seed being dipped for a time in a liquid containing the bacteria. In this manner organisms obtained from red clover can be grown and applied to the seed of red clover; and similar inoculation can be arranged for other species, so that an application of the bacteria most suited to the particular crop to be cultivated can be assured. In many cases it has been found that inoculation, whether of the soil or of the seed, has not made any appreciable difference to the growth of the crop, a result no doubt due to the fact that the soil had already contained within it an abundant supply of suitable organisms. But in other instances greatly increased yields have been obtained where inoculation has been practised.

More or less pure cultures of the nitrogen-fixing bacteria belonging to the *Azotobacter* group have been tried and recommended for application to poor land in order to provide a cheap supply of nitrogen. The application of pure cultures of bacteria for improving the fertility of the land is still in an experimental stage. There is little doubt, however, that in the near future means will be devised to obtain the most efficient work from these minute organisms, either by special artificial cultivation and subsequent application to the soil, or by improved methods of encouraging their healthy growth and activity in the land where they already exist.

IMPROVEMENT OF SOILS

The fertility of a soil is dependent upon a number of factors, some of which, such as the addition of fertilizers or manures, increase the stock of available food materials in the soil, while others, such as application of clay or humus, chiefly influence the fertility of the land by improving its physical texture.

The chief processes for the improvement of soils are liming, claying and marling, warping, paring and burning, and green manuring. Most of these more or less directly improve the land by adding to it certain plant food constituents which are lacking, but the effect of each process is in reality very complex. In the majority of cases the good results obtained are more particularly due to the setting free of " dormant " or

" latent " food constituents and to the amelioration of the texture of the soil, so that its aeration, drainage, temperature and water-holding capacity are altered for the better.

The material which chemists call calcium carbonate is met with in a comparatively pure state in chalk. It is present in variable amounts in limestones of all kinds, although its white ness may there be masked by the presence of iron oxide and other coloured substances. Carbonate of lime is also a constituent to a greater or lesser extent in almost all soils. In certain sandy soils and in a few stiff clays it may amount to less than 4%, while in others in limestone and chalk districts there may be 50 to 80% present.

Pure carbonate of lime when heated loses 44% of its weight, the decrease being due to the loss of carbon dioxide gas. The resulting white product is termed calcium oxide lime, burnt lime, quicklime, cob lime, or caustic lime. This substance absorbs and combines with water very greedily, at the same time becoming very hot, and falling into a fine dry powder,' calcium hydroxide or slaked lime, which when left in the open slowly combines with the carbon dioxide of the air and becomes calcium carbonate, from which we began.

When recommendations are made about liming land it is necessary to indicate more precisely than is usually done which of the three classes of material named above - chalk, quicklime or slaked lime - is intended. Generally speaking the oxide or quicklime has a more rapid and greater effect in modifying the soil than slaked lime, and this again greater than the carbonate or chalk.

Lime in whatever form it is applied has a many-sided influence in the fertility of the land. It tends to improve the tilth and the capillarity of the soil by binding sands together somewhat and by opening up clays. If applied in too great an amount to light soils and peat land it may do much damage by rendering them too loose and open.

The addition of small quantities of lime, especially in a caustic form, to stiff greasy clays makes them much more porous and pliable. A lump of clay, which if dried would become hard and intractable, crumbles into pieces when dried after adding to it 2% of lime. The lime causes the minute separate particles of clay to flocculate or group themselves together into larger compound grains between which air and water can percolate more freely. It is this power of creating a more crumbly tilth on stiff clays that makes lime so valuable to the farmer.

Lime also assists in the decomposition of the organic matter or humus in the soil and promotes nitrification; hence it is of great value after green manuring or where the land contains much humus from the addition of bulky manures such as farm-yard dung.

This tendency to destroy organic matter makes the repeated application of lime a pernicious practice, especially on land which contains little humus to begin with. The more or less dormant nitrogen and other constituents of the humus are made immediately available to the succeeding crop, but the capital of the soil is rapidly reduced, and unless the loss is replaced by the addition of more manures the land may become sterile.

Although good crops may follow the application of lime, the latter is not a direct fertilizer or manure and is no substitute for such. Its best use is obtained on land in good condition, but not where the soil is poor. When used on light dry land it tends to make the land drier, since it destroys the humus which so largely assists in keeping water in the soil. Lime is a base and neutralizes the acid materials present in badly drained meadows and boggy pastures. Weeds, therefore, which need sour conditions for development are checked by liming and the better grasses and clovers are encouraged. It also sets free potash and possibly other useful plant food-constituents of the soil. Liming tends to produce earlier crops and destroys the fungus which causes finger-and-toe or club-root among turnips and cabbages.

Land which contains less than about z/c, of lime usually needs the addition of this material. The particular form in which lime should be applied for the best results depends upon the nature of the soil. In practice the proximity to chalk pits or lime kilns, the cost of the lime and cartage, will determine which is most economical. Generally speaking light poor lands deficient in organic matter will need the less caustic form or chalk, while quicklime will be most satisfactory on the stiff clays and richer soils. On the stiff soils overl y ing the chalk it was formerly the custom to dig pits through the soil to the rock below. Shafts 20 or 30 ft. deep were then sunk, and the chalk taken from horizontal tunnels was brought to the surface and spread on the land at the rate of about 60 loads per acre. Chalk should be applied in autumn, so that it may be split by the action of frost during the winter.

Quicklime is best applied, perhaps, in spring at the rate of one ton per acre every six or eight years, or in larger doses- 4 to 8 tons - every 15 to 20 years. Small dressings applied at short intervals give the most satisfactory results. The quicklime should be placed in small heaps and covered with soil if possible until it is slacked and the lumps have fallen into powder, after which it may be spread and harrowed in. Experiments have shown that excellent effects can be obtained by applying 5 or 6 cwt. of ground quicklime.

Gas-lime is a product obtained from gasworks where quicklime is used to purify the gas from sulphur compounds and other objectionable materials. It contains a certain amount of unaltered caustic lime and slacked lime, along with sulphates and sulphides of lime, some of which have an evil odour. As some of these sulphur compounds have a poisonous effect on plants, gas-lime cannot be applied to land directly without great risk or rendering it incapable of growing crops of any sort - even weeds - for some time. It should therefore be kept a year or more in heaps in some waste corner and turned over once or twice so that the air can gain access to it and oxidize the poisonous ingredients in it.

Many soils of a light sandy or gravelly or peaty nature and liable to drought and looseness of texture can be improved by the addition of large amounts of clay of an ordinary character. Similarly soils can be improved by applying to them marl, a substance consisting of a mixture of clay with variable proportions of lime. Some of the chalk marls, which are usually of a yellowish or dirty grey colour, contain clay and 50 to 80%

of carbonate of lime with a certain proportion of phosphate of lime. Such a material would not only have an influence on the texture of the land but the lime would reduce the sourness of the land and the phosphate of lime supply one of the most valuable of plant foodconstituents. The beneficial effects of marls may also be partially due to the presence in them of available potash.

Typical clay-marls are tenacious, soapy clays of yellowish-red or brownish colour and generally contain less than 50% of lime. When dry they crumble into small pieces which can be readily mixed with the soil by ploughing. Many other kinds of marls are described; some are of a sandy nature, others stony or full of the remains of small shells. The amount and nature of the clay or marl to be added to the soil will depend largely upon the original composition of the latter, the lighter sands and gravel requiring more clay than those of firmer texture. Even stiff soils deficient in lime are greatly improved in fertility by the addition of marls.

In some cases as little as 40 loads per acre have been used with benefit, in others 180 loads have not been too much. The material is dug from neighbouring pits or sometimes from the fields which are to be improved, and applied in autumn and winter. When dry and in a crumbly state it is harrowed and spread and finally ploughed in and mixed with the soil.

On some of the strongest land it was formerly the practice to add to and plough into it burnt clay, with the object of making the land work more easily. The burnt clay moreover carried *Cl ay* with it potash and other materials in a state readily available to the crops. The clay is dug from the land or from ditches or pits and placed in heaps of 60 to ioo loads each, with faggot wood, refuse coals or other fuel. Great care is necessary to prevent the heaps from becoming too hot, in which case the clay becomes baked into hard lumps of brick-like material which cannot be broken up. With careful management, however, the clay dries and bakes, becoming slowly converted into lumps which readily crumble into a fine powder, in which state it is spread over and worked into the land at the rate of 40 loads per acre.

The paring and burning of land, although formerly practised as an ordinary means of improving the texture and fertility of arable fields, can now only be looked upon as a practice p to be adopted for the purpose of bringing rapidly into cultivation very foul leys or, land covered with a coarse turf. The practice is confined to poorer types of land, such as heaths covered with furze and bracken or fens and clay areas smothered with rank grasses and sedges.

To reduce such land to a fit state for the growth of arable crops is very difficult and slow without resort to paring and burning. The operation consists of paring off the tough sward to a depth of I to 2 in. just sufficient to effectually damage the roots of the plants forming the sward and then, after drying the sods and burning them, spreading the charred material and ashes over the land. The turf is taken off either with the breast plough - a paring tool pushed forward from the breast or thighs by the workman - or with specially constructed paring ploughs or shims. The depth of the sod removed should not be too thick or burning is difficult and too much humus is destroyed unnecessarily, nor should it be too thin or the roots of the herbage are not effectually destroyed.

The operation is best carried out in spring and summer. After being pared off the turf is allowed to dry for a fortnight or so and is then placed in small heaps a yard or two wide at the base, a little straw or wood being put in the middle of each heap, which is then lighted.

As burning proceeds more turf is added to the outside of the heaps in such a manner as to allow little access of air. Every care should be taken to burn and char the sod thoroughly without permitting the heap to blaze. The ashes should be spread as soon as possible and covered by a shallow ploughing. The land is then usually sown with some rapidly growing green crop, such as rape, or with turnips.

Paring and burning improves the texture of clay lands, particularly if draining is carried out at the same time. It tends to destroy insects and weeds, and gets rid of acidity of the soil. No operation brings old turf into cultivation so rapidly. Moreover the beneficial effects are seen in the first crop and last for many years. Many of the mineral plant food-constituents locked up in the coarse herbage and in the upper layers of the soil are made immediately available to crops. The chief disadvantage is the loss of nitrogen which it entails, this element being given off into the air in a free gaseous state. It is best adapted for application to clays and fen lands and should not be practised on shallow light sands or gravelly soils, since the humus so necessary for the fertility of such areas is reduced too much and the soil rendered too porous and liable to suffer from drought.

Many thousands of acres of low-lying peaty and sandy land adjoining the tidal rivers which flow into the Humber have been improved by a process termed " warping." The warp consists of fine muddy sediment which is suspended in the tidal river water and appears to be derived from material scoured from the bed of the Humber by the action of the tide and a certain amount of sediment brought down by the tributary streams which join the Humber some distance from its mouth. The field or area to be warped must lie below the level of the water in the river at high tide.

It is first surrounded by an embankment, after which the water from the river is allowed to flow through a properly constructed sluice in its bank, along a drain or ditch to the land which is prepared for warping. By a system of carefully laid channels the water flows gently over the land, and deposits its warp with an even level surface. At the ebb of the tide the more or less clear water flows back again from the land into the main river with sufficient force to clean out any deposit which may have accumulated in the drain leading to the warped area, thus allowing free access of more warpladen water at the next tide. In this manner poor peats and sands may be covered with a large layer of rock soil capable of growing excellent crops.

The amount of deposit laid over the land reaches a thickness of two or three feet in one season of warping, which is usually practised between March and October, advantage being taken of the spring tides during these months. The new warp is allowed to lie fallow during the winter after being laid out in four-yard " lands " and becomes dry enough to be sown with oats and grass and clover seeds in the following spring. The clover-grass ley is then grazed for a year or two with sheep, after which wheat and potatoes are the chief crops grown on the land.

Green manures are crops which are grown especially for the purpose of ploughing into the land in a green or actively growing state. The crop during its growth obtains a considerable amount of *Green.* carbon from the carbon dioxide of the air, and builds it up *Manuring. i* nto compounds which when ploughed into the land become humus. The carbon compounds of the latter are of no direct nutritive value to the succeeding crop, but the decaying vegetable tissues very greatly assist in retaining moisture in light sandy soils, and in clay soils also have a beneficial effect in rendering them more open and allowing of better drainage of superfluous water and good circulation of fresh air within them.

The ploughing-in of green crops is in many respects like the addition of farm-yard manure. Their growth makes no new addition of mineral food-constituents to the land, but they bring useful substances from the subsoil nearer to the surface, and after the decay of the buried vegetation these become available to succeeding crops of wheat or other plants. Moreover, where deep-rooting plants are grown the subsoil is aerated and rendered more open and suitable for the development of future crops.

The plants most frequently used are white mustard, rape, buckwheat, spurry, rye, and several kinds of leguminous plants, especially vetches, lupins and serradella. By far the most satisfactory crops as green manures are those of the leguminous class, since they add to the land considerable amounts of the valuable fertilizing constituent, nitrogen, which is obtained from the atmosphere.

By nitrification this substance rapidly becomes available to succeeding crops. On the light, poor sands of Saxony Herr Schultz, of Lupitz, made use of serradella, yellow lupins and vetches as green manures for enriching the land in humus and nitrogen, and found the addition of potash salts and phosphates very profitable for the subsequent growth of potatoes and wheat.

He estimated that by using leguminous crops in this manner for the purpose of obtaining cheap nitrogen he reduced the cost of production of wheat more than 50%. The growing crops should be ploughed in before flowering occurs; they should not be buried deeply, since decay and nitrification take place most rapidly and satisfactorily when there is free access of air to the decaying material. When the crop is luxuriant it is necessary to put a roller over it first, to facilitate proper burial by the plough. The best time for the operation appears to be late summer and autumn.

Soil and Disease

The influence of different kinds of soil as a factor in the production of disease requires to be considered, in regard not only to the nature and number of the microorganisms they contain, but also to the amount of moisture and air in them and their capacity for heat. The moisture in soil is derived from two sources—the rain and the ground-water.

Above the level of the ground-water the soil is kept moist by capillary attraction and by evaporation of the water below, by rainfall, and by movements of the ground-water; on the other hand, the upper layers are constantly losing moisture by evaporation from the surface and through vegetation. When the ground-water rises it forces air out of the soil; when it falls again it leaves the soil moist and full of air. The nature of the soil will largely influence the amount of moisture which it will take up or retain. In regard to water, all soils have two actions - namely, permeability and absorbability.

Permeability is practically identical with the speed at which percolation takes place; through clay it is slow, but increases in rapidity through marls, loams, limestones, chalks, coarse gravels and fine sands, reaching a maximum in soil saturated with moisture.

The amount of moisture retained depends mainly upon the absorbability of the soil, and as it depends largely on capillary action it varies with the coarseness or fineness of the pores of the soil, being greater for soils which consist of fine particles.

The results of many analyses show that the capacity of soils for moisture increases with the amount of organic substances present; decomposition appears to be most active when the moisture is about 4%, but can continue when it is as low as 2%, while it appears to be retarded by any excess over 4%. Above the level of the ground-water all soils contain air, varying in amount with the degree of looseness of the soil. Some sands contain as much as 50% of air of nearly the same composition as atmospheric air. The oxygen, however, decreases with the depth, while the carbon dioxide increases.

Among the most noteworthy workers at the problems involved in the question of the influence of soil in the production of disease we find yen Foder, Pettenkofer, Levy, Fleck, von Naegeli, Schleesing, Muntz and Warrington. The study of epidemic and endemic diseases generally has brought to light an array of facts which very strongly suggest that an intimate association exists between the soil and the appearance and propagation of certain diseases; but although experiments and observations allow this view to be looked upon as well established, still the precise role played by the soil in an aetiological respect is by no means so well understood as to make it possible to separate the factors and dogmatize on their effects.

The earliest writers upon cholera emphasized its remarkable preference for particular places; and the history of each successive epidemic implies, besides an importation of the contagion, certain local conditions which may be either general sanitary defects or peculiarities of climate and soil.

The general evidence indicates that the specific bacteria of cholera discharges are capable of a much longer existence in the superficial soil layers than was formerly supposed; consequently it is specially necessary to guard against pollution of the soil, and through it against the probable contamination of both water and air.

The evidence, however, is not sufficiently strong to warrant a universal conclusion, the diffusion of cholera appearing to be largely dependent upon other factors than soil states. Again, all accounts of diphtheria show a tendency on the part of the disease to recur in the same districts year after year. The questions naturally suggest themselves - Are the reappearances due to a revival of the contagion derived from previous outbreaks in the same place, or to some favouring condition which the place offers for the development of infection derived from some other quarter; and have favouring conditions any dependence upon the character and state of the soil? Greenhow in 1858 stated that diphtheria was especially prevalent on cold, wet soils, and Airy in 1881 described the localities affected as " for the most part cold, wet, clay lands."

An analysis of the innumerable outbreaks in various parts of Europe indicates that the geological features of the affected districts play a less important part in the incidence of the disease than soil dampness. In this connexion it is interesting to note the behaviour of the diphtheritic contagion in soil.

Experiments show that pure cultures, when mixed with garden soil constantly moistened short of saturation and kept in the dark at a temperature of 14° C., will retain their vitality for more than ten months; from moist soil kept at 26° C. they die out in about two months; from moist soil at 30° C. in seventeen days; and in dry soil at the same temperature within a week. In the laboratory absolute soil dryness is as distinctly antagonistic to the vitality of the diphtheria bacillus as soil dampness is favourable.

Both statistically and experimentally we find that a damp soil favours its life and development, while prolonged submersion and drought kill it. We may consider that, in country districts, constant soil moisture is one of the chief factors; while in the case of urban outbreaks mere soil moisture is subsidiary to other more potent causes.

Again, many facts in the occurrence and diffusion of enteric fever point to an intimate connexion between its origin and certain conditions of locality. Epidemics rarely spread over any considerable tract of country, but are nearly always confined within local limits. Observations made at the most diverse parts of the globe, and the general distribution area of the disease, show that mere questions of elevation, or even configuration of the ground, have little or no influence.

On the other hand, the same observations go to show that the disease is met with oftener on the more recent formations than the older, and this fact, so far as concerns the physical characters of the soil, is identical with the questions of permeability to air and water. Robertson has shown that the typhoid bacillus can grow very easily in certain soils, can persist in soils through the winter months, and when the soil is artificially fed, as may be done by a leaky drain or by access of filthy water from the surface, the microorganism will take on a fresh growth in the warm season. The destructive power of sunlight is only exercised on those organisms actually at the surface.

Cultures of the typhoid organism planted at a depth of 18 in. were found to have grown to the surface. In the winter months the deeper layers of the soil act as a shelter to the organism, which again grows towards the surface during the summer. The typhoid organism was not found to be taken off from the decomposing masses of semi-liquid filth largely contaminated with a culture of *bacillus typhosus;* but, on the other hand, it was abundantly proved that it could grow over moist surfaces of stones, &c.

Certain disease-producing organisms, such as the bacillus of tetanus and malignant oedema, appear to be universally distributed in soil, while others, as the *bacillus typhosus* and *spirillum cholerae,* appear to have only a local distribution. The conditions which favour the vitality, growth and multiplication of the typhoid bacillus are the following: the soil should be pervious; it should be permeated with a sufficiency of decaying - preferably animal - organic matters; it should possess a certain amount of moisture, and be subject to a certain temperature. Depriving the organism of any of these essential conditions for its existence in the soil will secure our best weapon for defence.

The optimum temperature adapted to its growth and extension is 37° C. =98°. 4 F. Sir Charles Cameron attributes the prevalence of typhoid in certain areas in Dublin to the soil becoming saturated with faecal matter and specifically infected.

The ratio of cases to population living in Dublin on loose porous gravel soil for the ten years1881-1891was I in 94, while that of those living on stiff clay soil was but 1 in 145. " This is as we should expect, since the movements of ground air are much greater in loose porous soils than in stiff clay soils." A foul gravel soil is a most dangerous one on which to build.

For warmth, for dryness, for absence of fog, and for facility of walking after rain, just when the air is at its purest and its best, there is nothing equal to gravel; but when gravel has been rendered foul by infiltration with organic matters it may easily become a very hotbed of disease.

Chapter 6

Stellar Evolution

All stars form from clouds of gas and dust condensing in deep space. Only the chemical composition of this cloud, and the amount of material in the cloud that condenses into the actual star, determines what will happen to the star over its entire lifetime. As an interstellar gas cloud starts to condense under its own gravitation, any tiny amount of spin that it has will become amplified, the way a whirling figure skater spins faster when he brings in his arms. Eventually, little whirlpools or eddies will form in this ever-more-rapidly-spinning collapsing cloud. It's these eddies that will eventually form star systems. All that gaseous material falling in on itself in a given eddy releases an enourmous amount of heat when it starts to collide with itself. The more the whirlpool contracts, the hotter and more opaque it gets, until it gets hot enough and thick enough to glow. Such an object is called a *protostar*; we can see such an object from here on Earth, provided the cloud of gas and dust surrounding it is thin enough to see through.

When the protostar is nearly finished collapsing under its own weight, it will reach its maximum temperature. On the surface, it will actually be hotter than it will when it becomes a main-sequence star. But it's the temperature deep within its core that determines the protostar's fate. In most cases, the protostar's total mass will be less than about eight percent the mass of the sun, and the core temperature and pressure will not be high enough for thermonuclear reactions to begin; or, if they are, the initial belch of nuclear activity will push the outer layers of the protostar outward and rarefy the core enough to snuff the fusion reactions out. Such a "failed star" is called a brown dwarf and is probably one of the most plentiful, if hard-to-detect, objects in the galaxy. In some cases, though, the protostar's mass (and therefore its peak core temperature) will be high enough to ignite stable thermonuclear reactions. Soon thereafter, the fusion energy released from the new stellar core reaches its surface, the initial birthing contractions finish, and the newborn star settles down onto the Main Sequence, where it will spend most of its productive lifetime.

Since main-sequence stars do not shrink appreciably over time, *all* of a main-sequence star's radiant energy must be produced in the core by hydrogen fusion. There are two distinct types of hydrogen-burning reactions that stellar core material can undergo. Main-sequence stars lighter than about class F0 fuse hydrogen into helium via the *proton-proton chain*. This is a rather straightforward nuclear reaction:

- Two protons fuse together, forming a deuterium nucleus and releasing both a neutrino and a positron (the positron eventually annihilates with an electron to produce energy);
- Then, another proton collides with the deuterium nucleus, forming a helium-3 nucleus and giving off a gamma ray photon;
- Finally, another helium-3 nucleus formed by steps 1 and 2 above collides with this helium-3 nucleus, turning it into an ordinary helium-4 nucleus and releasing two protons. The total reaction time for this entire process is on the order of one million years.

Heavier main-sequence stars take advantage of their higher core temperatures to fuse hydrogen into helium more rapidly, by a process called the *CNO cycle*. This is a six-step process which uses ordinary carbon-12 as a kind of nuclear catalyst. The net result is the same: four protons turn into a helium-4 nucleus and two positrons, liberating energy in the process, while all the other materials that partook in the reaction come out unchanged. (as carbon is required for this reaction, galactic halo population stars will be too heavy-element-poor to undergo it on a large scale; heavy main-sequence stars in the galactic halo use the proton-proton chain just like lighter stars do.) Unlike the slow proton-proton chain, a CNO cycle reaction is about a thousand times faster, taking only a thousand or so years to complete. This means that heavier main-sequence stars that are heavy-element-rich will shine much more brightly than lighter main-sequence stars. It also means that the heavier stars will burn out their core's supply of nuclear fuel much faster.

The luminosities and estimated main-sequence lifetime for stars hotter than spectral class F0 assumes it is heavy-element-rich enough to have sufficient carbon to run the CNO cycle; a heavy-element-poor star hotter than F0 would be considerably dimmer and last considerably longer. It should also be noted that the currently estimated age of the universe, according to big bang theory, is between 10 000 and 20 000 million years — shorter than the lifespan of a class K or M main-sequence star. This means it should be impossible to find the remnants of any former K or M main-sequence stars anywhere in the known universe. If we ever find any, our picture of the universe — or of stellar evolution — will have to be revised.)

And what happens to a star when it's reached the end of its main sequence lifetime, when it's exhausted about half the available fuel in its core and can no longer sustain a hydrogen fusion reaction at the rate it once did? Well, like its properties during its main sequence lifetime, that all depends on the mass of the star.

LIGHTWEIGHT STARS

Stars whose main sequence spectral class was anywhere from M on up through the A's will start the Beginning of the End by slowly expanding into a Red Giant. When nuclear fuel is no longer plentiful in the core, it can no longer maintain its main-sequence outward pressure and begins to contract under its own weight. As it collapses, the layers above it fall inward on top of it, causing them to heat up. Soon, the layer immediately above the core will become hot enough and high-pressure enough to undergo thermonuclear reactions on its own – and since this layer has an ample supply of hydrogen (unlike the exhausted core), it becomes a self-sustaining *hydrogen-burning shell* and will actually burn hydrogen into helium faster than the core did during its main-sequence lifetime. The added energy and outward pressure from this hydrogen-burning shell stops the collapse of the upper layers; in fact they begin expanding, and will keep expanding until the star becomes a Red Giant. It takes thousands of years for a star to grow from initial-collapse-at-the-end-of-the-main-sequence to the full-blown red giant stage (a 1962 study claims that it takes "only" about 20 000 years for a spctral class A main-sequence star to evolve into a class M red giant).

After a few million years, the new hydrogen-burning shell will exhaust itself also. This causes the star to contract under its own weight once again. Briefly, the super-compacted core may flash into life, fusing helium into carbon for a brief instant measured literally in seconds (the reaction rate for helium fusion is about a million million times faster than hydrogen fusion), but as helium-fusion produces much less energy than hydrogen-fusion does, and since the core is buried so deeply within the star, this *helium flash* will not be seen and is only predicted in theory. Finally, as this last burp of energy generated by the helium flash slowly reaches the surface, the star becomes a red giant a second time, sheds up to half its mass into interstellar space as a so-called "planetary nebula," and leaves only its core behind.

The core that it leaves behind, though, is a fascinating object. It weighs about half of what the star did during its main sequence lifetime, yet it's smaller than Uranus or Neptune. It's hotter than the star was when it was on the main sequence, and gives off blackbody radiation just like a hot star would; yet it produces no energy of its own and glows simply because it hasn't cooled off yet.

Its surface gravity can measure well over 100 000 times the surface gravity of the Earth. Its average density is over a ton to the cubic centimeter; it is so incredibly dense, in fact, that all the atoms that make it up are packed together as tightly as the laws of Fermion physics will allow, making it a totally incompressible "electron degenerate" gas. This oddball superdense stellar remnant is called a *white dwarf.*

Electron-degeneracy theory predicts that the uppermost mass a white dwarf can attain is about 1.4 times the mass of the sun, called the Chandrasekhar Limit. Any heavier, and the tremendous pressure on the innermost atoms would squeeze their electrons into the nuclei they orbit, turning all the protons and electrons in the star into neutrons. So far, no white dwarfs of more than 1.4 solar masses have been found, so the theory seems to be on firm ground.

The low surface area and high specific heat of a white dwarf means that such an object would take a *long* time to cool off — longer, even, than the currently estimated age of the universe. If the universe were a few hundred thousand million years older, we would expect it to be populated by white dwarfs that have cooled off below the point where they glow; these academic objects are referred to as *black dwarfs*.

MIDDLE WEIGHT STARS

A class B main-sequence star will leave the main sequence much as lighter stars do, collapsing a little, forming a hydrogen-burning shell, turning into a Red Giant (or a Cepheid variable like Polaris), shrinking again as its hydrogen-burning shell exhausts itself, then shining more brightly as its core goes through a helium-burning phase. The difference now, though, is that burning helium into carbon in the star's core is no longer the end of the road. As this fuel supply runs out, the star's collapse reignites the depleted hydrogen-burning shell and turns it into a *helium-burning shell*. This renewed energy then creates a new hydrogen burning shell in a layer above the old one, so that as we move inward from the star's surface, we get a hydrogen burning shell, then a helium burning shell, and finally the core underneath. The core will likewise undergo renewed thermonuclear vigour, fusing its old carbon together with more helium to form oxygen.

When this stage completes, the core can begin fusing oxygen into neon, the old helium-burning shell can become a carbon-burning shell, the formerly outermost hydrogen-burning shell becomes the new helium-burning shell, and yet another thin hydrogen-burning shell emerges outside of that. And then, neon can fuse into magnesium, then magnesium can fuse into silicon, and so on down the periodic chart until, finally, chromium gets fused into iron. Each of these fusion stages (helium-to-carbon, carbon-to-oxygen,. . ., chromium-to-iron) produces less energy than the preceding stage does, and thus exhausts its own fuel supply ever more rapidly. During these late stages of its evolution, the star can bloat up to hundreds of times the diameter of the sun, becoming a red supergiant like Betelgeuse.

Finally, though, when the star gets around to wanting to fuse iron into something heavier, it runs into a problem. Iron is at the "bottom of the well" when it comes to nuclear reactions. Fusing it into something heavier, or for that matter breaking it apart into something lighter, always consumes more energy than it produces. So when the core starts to "burn" iron, it ends up getting cooler, not hotter. All the outward pressure that its nuclear reactions have been generating suddenly vanishes.

The star's core collapses in the blink of an eye. And, since the core takes up such a large fraction of the star's total mass, it's heavier than the maximum 1.4 solar masses that can support a white dwarf. Its protons and electrons get squeezed together until it is a solid ball of neutrons, no bigger across than Los Angeles and with the density of an atomic nucleus (around a thousand million tons to the cubic centimeter). It is now a *neutron star* and is said to be "neutron-degenerate." The surface gravity of such a beast is on the order of a million million G's.

In collapsing in on itself to such dense proportions, all of the core's gravitational potential energy has to be released in the form of heat, just like the collapsing cloud that originally formed the star heated up as it contracted. This time, though, the amount of energy released is much greater and happens over the span of a few seconds. All the outer layers of the star, even those that never became nuclear fusion shells, will become superheated plasmas hot enough to fuse their constituent ions into not only iron, but copper, strontium, silver, gold, lead — even uranium. These super-hot, super-bright outer layers race off into interstellar space at nearly the speed of light, carrying their newly-formed heavy elements with them and creating one of the most spectacular and rare sights in the heavens: a type II supernova.

(Incidentally, it's believed that supernovae are the *only* phenomena that can send heavy elements into the interstellar medium. Thus, the heavy element enrichment that our solar system enjoys is thought to be the product of earlier supernovas that infused their products into the cloud that our own sun (and its planets) condensed out of.)

With the aid of telescopes, the expanding cloud from a type II supernova can be seen for many centuries hence as a nebula (such as the crab nebula). The neutron star left at the cloud's centre is too small to be seen with current instruments, but it can be detected by its radio emissions *if* one of its magnetic poles happens to sweep past the Earth as the star rotates. (Its collapse to such a compact object means it will be spinning very rapidly; its magnetic pole may sweep past the Earth hundreds of times a second. It would thus appear to a radio telescope to be a very rapidly, regularly pulsating radio source called a *pulsar*.)

HEAVYWEIGHT STARS

The rare class O main-sequence stars start the end of their lives as the middleweight stars do, bloating, forming energy-producing shells around the core, and fusing heavier and heavier elements together until the core becomes iron. And, once again, when the core attempts to fuse iron into something heavier, it loses its energy support and collapses, crossing the Chandrasekhar Limit and squeezing itself into a ball of neutrons.

There is, however, a theoretical limit on how heavy even a neutron star can become. Past about three solar masses, even neutron degeneracy can't support the core's weight. In fact, no force known can support its weight. The core continues to collapse until it is an infinitely small, infinitely dense point called a *singularity*. Its gravity will be so strong that neither the material from the original core's outer layers, nor the energy from the core's collapse, nor even a beam of light directed straight outward can escape it. *Nothing* that comes within the Schwarzchild Radius (3 kilometers times the mass of the singularity in solar masses) can escape it.

As far as the outer layers of the star are concerned, the core has merely fizzled out, removing its energy support and letting them fall; these outer layers too will fall within the singularity's gravitational grip never to be seen again. The whole star swallows itself, leaving only its gravity behind; it's now called a *black hole*.

A word about Novae and X-ray Bursters

In a binary star system, one star will usually be more massive than the other, meaning that the heavier star of the pair may end its main-sequence lifetime millions or thousands of millions of years before its lighter companion does. Many white dwarfs, for instance, have been detected because of oddities in the movement or appearance of their main-sequence companion Sirius B being the most famous example, discovered by accident over a century ago when a new telescope lens resolved Sirius's companion during a test. Sometimes, due to orbital decay or the fact that the longer-lived companion star has reached the end of its lifetime and is turning into a red giant, a white dwarf, neutron star, or black hole can come so close to its binary host-star that its strong gravity begins drawing (or "accreting") material off its host. Such a system is called a *mass-exchange binary*.

This sucked-up gas swirls around the white dwarf or neutron star, forming an *accretion disk* as it siprals in toward its new owner's surface. In the case of a neutron star or black hole, the accretion disk will be the only feature of the companion star visible from the Earth. Material accreted "onto" a black hole essentially goes down the event-horizon drain and is gone forever. Material accreted onto a neutron star or white dwarf, however, will accumulate on that star's surface, forming thicker and thicker layers of supercompressed hydrogen. If the infalling material is moving fast enough, this accreted hydrogen can gain sufficient heat and pressure for thermonuclear reactions to occur.

Depending on how fast the incoming material is moving, several things can happen to a white dwarf. Very rapidly infalling material will ignite all at once, causing the white dwarf to shine several times more brightly than its companion for a few days, then taper off back down to its original brightness. Years or centuries later, the process may repeat itself. This phenomenon is called a *nova* (the Latin word for "new") because, to the unaided eye, it looks like a new star has appeared in the sky where before there was none. If the accreted material is trickling in more slowly, it will only ignite in small spurts, turning the white dwarf into a *cataclysmic variable*. If the accreted material accumulates *very* slowly, the white dwarf can heat up as a whole, until the entire star blows itself apart in one massive thermonuclear fireball called a type I supernova.

A neutron star whose accreted layers ignite will burn all its available newfound hydrogen into helium in a matter of seconds. This is only visible as an intense burst of X-rays lasting for, at most, a minute or two. The process repeats itself sporadically every few hours as new material is accreted. Not surprisingly, such phenomena are called *X-ray bursters*.

STELLAR REMNANTS

After a star has burned out its fuel supply, its remnants can take one of three forms, depending on the mass during its lifetime.

White Dwarfs

For a star of 1 solar mass, the resulting white dwarf is of about 0.6 solar masses, compressed into approximately the volume of the Earth. White dwarfs are stable because the inward pull of gravity is balanced by the degeneracy pressure of the star's electrons. Electron degeneracy pressure provides a rather soft limit against further compression; therefore, for a given chemical composition, white dwarfs of higher mass have a smaller volume. With no fuel left to burn, the star radiates its remaining heat into space for billions of years.

The chemical composition of the white dwarf depends upon its mass. A star of a few solar masses will ignite carbon fusion to form magnesium, neon, and smaller amounts of other elements, resulting in a white dwarf composed chiefly of oxygen, neon, and magnesium, provided that it can lose enough mass to get below the Chandrasekhar limit, and provided that the ignition of carbon is not so violent as to blow apart the star in a supernova. A star of mass on the order of magnitude of the Sun will be unable to ignite carbon fusion, and will produce a white dwarf composed chiefly of carbon and oxygen, and of mass too low to collapse unless matter is added to it later. A star of less than about half the mass of the Sun will be unable to ignite helium fusion and will produce a white dwarf composed chiefly of helium.

In the end, all that remains is a cold dark mass sometimes called a black dwarf. However, the universe is not old enough for any black dwarf stars to exist yet.

If the white dwarf's mass increases above the Chandrasekhar limit, which is 1.4 solar masses for a white dwarf composed chiefly of carbon, oxygen, neon, and/or magnesium, then electron degeneracy pressure fails due to electron capture and the star collapses. Depending upon the chemical composition and pre-collapse temperature in the centre, this will either lead to collapse into a neutron star or runaway ignition of carbon and oxygen. Heavier elements favour continued core collapse, because they require a higher temperature to ignite, because electron capture onto these elements and their fusion products is easier; higher core temperatures favour runaway nuclear reaction, which halts core collapse and leads to a Type Ia supernova. These supernovae may be many times brighter than the Type II supernova marking the death of a massive star, even though the latter has the greater total energy release. This instability to collapse means that no white dwarf more massive than approximately 1.4 solar masses can exist (with a possible minor exception for very rapidly spinning white dwarfs, whose centrifugal force due to rotation partially counteracts the weight of their matter). Mass transfer in a binary system may cause an initially stable white dwarf to surpass the Chandrasekhar limit. If a white dwarf forms a close binary system with another star, hydrogen from the larger companion may accrete around and onto a white dwarf until it gets hot enough to fuse in a runaway reaction at its surface, although the white dwarf remains below the Chandrasekhar limit. Such an explosion is termed a nova.

Neutron Stars

When a stellar core collapses, the pressure causes electron capture, thus converting the great majority of the protons into neutrons. The electromagnetic forces keeping separate nuclei apart are gone (proportionally, if nuclei were the size of dust motes, atoms would be as large as football stadiums), and most of the core of the star becomes a dense ball of contiguous neutrons (in some ways like a giant atomic nucleus), with a thin overlying layer of degenerate matter (chiefly iron unless matter of different composition is added later). The neutrons resist further compression by the Pauli Exclusion Principle, in a way analogous to electron degeneracy pressure, but stronger.

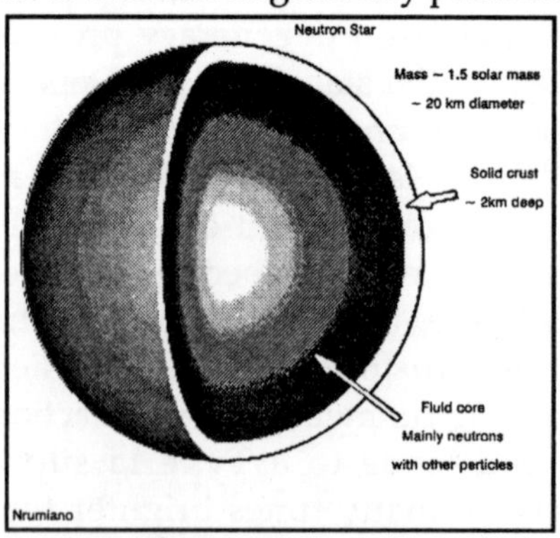

Fig. Neutron Stars

These stars, known as neutron stars, are extremely small on the order of radius 10 km, no bigger than the size of a large city and are phenomenally dense. Their period of revolution shortens dramatically as the star shrinks (due to conservation of angular momentum); some spin at over 600 revolutions per second. When these rapidly rotating stars' magnetic poles are aligned with the Earth, a pulse of radiation is received each revolution. Such neutron stars are called pulsars, and were the first neutron stars to be discovered.

Black Holes

If the mass of the stellar remnant is high enough, the neutron degeneracy pressure will be insufficient to prevent collapse below the Schwarzschild radius. The stellar remnant thus becomes a black hole. The mass at which this occurs is not known with certainty, but is currently estimated at between 2 and 3 solar masses.Black holes are predicted by the theory of general relativity. According to classical general relativity, no matter or information can flow from the interior of a black hole to an outside observer, although quantum effects may allow deviations from this strict rule. The existence of black holes in the universe is well supported, both theoretically and by astronomical observation.

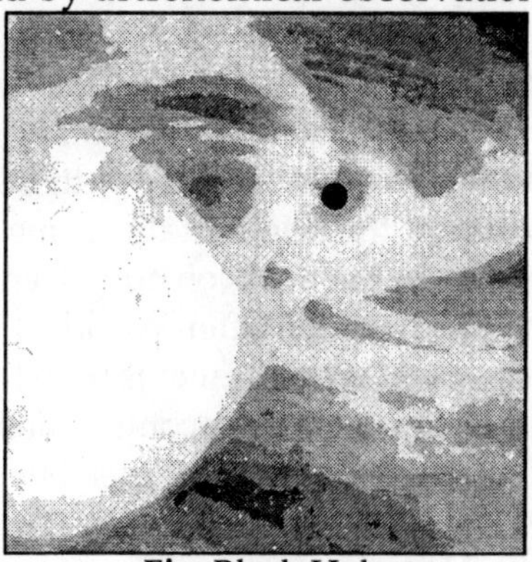

Fig. Black Hole

Since the core-collapse supernova mechanism itself is imperfectly understood, it is still not known whether it is possible for a star to collapse directly to a black hole without producing a visible supernova, or whether some supernovae initially form unstable neutron stars which then collapse into black holes; the exact relation between the initial mass of the star and the final remnant is also not completely certain.

Resolution of these uncertainties requires the analysis of more supernovae and supernova remnants.

NUCLEOSYNTHESIS

Nucleosynthesis is the process of creating new atomic nuclei from preexisting nucleons (protons and neutrons). The primordial nucleons themselves were formed from the quark-gluon plasma of the Big Bang as it cooled below ten million degrees. A few minutes afterward, starting with only protons and neutrons, nuclei up to lithium and beryllium (both with mass number 7) were formed but only in relatively small amounts. This first process of primordial nucleosynthesis may also be called nucleogenesis. The subsequent nucleosynthesis of the elements (including all carbon, all oxygen, etc.) occurs primarily in stars either by nuclear fusion or nuclear fission.

The first ideas were that the chemical elements were created at the beginnings of the universe, but no successful picture could be found. Arthur Stanley Eddington first suggested in 1920 that stars obtain their energy by fusing hydrogen to helium, but this idea was not generally accepted because it lacked nuclear mechanisms. In the years immediately before World War II Hans Bethe first provided those nuclear mechanisms by which hydrogen is fused into helium. However, neither of these early works on stellar power addressed the origin of the elements heavier than helium. Fred Hoyle's original work on nucleosynthesis of heavier elements in stars occurred just after World War II. This work attributed production of heavier elements from hydrogen in stars during the nuclear evolution of their compositions. Hoyle's work explained how the abundances of the elements increased with time as the galaxy aged.

Subsequently, Hoyle's picture was expanded during the 1960s by creative contributions by William A. Fowler, Alistair G. W. Cameron, and Donald D. Clayton, and then by many others. The creative 1957 review paper by E. M. Burbidge, G. R. Burbidge, Fowler and Hoyle is a well-known summary of the state of the field in 1957. That paper defined new processes for changing one heavy nucleus into others within individual stars, processes that could be documented by astronomers.

Processes

There are a number of astrophysical processes which are believed to be responsible for nucleosynthesis in the universe. The majority of these occur within the hot matter inside stars. The successive nuclear fusion processes which occur inside stars are known as hydrogen burning (via the proton-proton chain or the CNO cycle), helium burning, carbon burning, neon burning, oxygen burning and silicon burning.

These processes are able to create elements up to iron and nickel, the region of the isotopes having the highest binding energy per nucleon. Heavier elements can be assembled within stars by a neutron capture process known as the s process or in explosive environments, such as supernovae, by a number of processes. Some of the more important of these include the r process which involves rapid neutron captures, the rp process which involves rapid proton captures and the p process (sometimes known as the gamma process) which involves photodisintegration of existing nuclei.

THE FOUR MAJOR TYPES OF NUCLEOSYNTHESIS

Big Bang Nucleosynthesis

Chief nuclear reactions responsible for the relative abundances of light atomic nuclei observed throughout the universe. Big Bang nucleosynthesis occurred within the first three minutes of the universe and is responsible for much of the abundance ratios of ^1H (protium), ^2H (deuterium), ^3He (helium-3), and ^4He (helium-4), in the universe.

$$n^0 \longrightarrow p^+ + e^- + \bar{\nu}_e \qquad p^+ + n^0 \longrightarrow {}^2_1D + \gamma$$

$${}^2_1D + p^0 \longrightarrow {}^3_2He + \gamma \qquad {}^2_1D + {}^2_1D \longrightarrow {}^3_2He + n^0$$

$${}^2_1D + {}^2_1D \longrightarrow {}^3_1T + P^+ \qquad {}^3_1T + {}^2_1D \longrightarrow {}^4_2He + n^0$$

$${}^3_2He + {}^4_2He \longrightarrow {}^7_3Li + \gamma \qquad {}^3_2He + n^0 \longrightarrow {}^3_1T + p^+$$

$${}^3_2He + {}^2_1D \longrightarrow {}^4_2He + p^+ \qquad {}^3_2He + {}^4_2He \longrightarrow {}^7_4Be + \gamma$$

$${}^7_3Li + p^+ \longrightarrow {}^4_2He + {}^4_2He \qquad {}^7_4Be + n^0 \longrightarrow {}^7_3Li + p^+$$

Although ^4He continues to be produced by other

mechanisms (such as stellar fusion and alpha decay) and trace amounts of ^{1}H continue to be produced by spallation and certain types of radioactive decay (proton emission and neutron emission), most of the mass of these isotopes in the universe, and all but the insignificant traces of the ^{3}He and deuterium in the universe produced by rare processes such as cluster decay, are thought to have been produced in the Big Bang.

The nuclei of these elements, along with some ^{7}Li, and ^{7}Be are believed to have been formed when the universe was between 100 and 300 seconds old, after the primordial quark-gluon plasma froze out to form protons and neutrons. Because of the very short period in which Big Bang nucleosynthesis occurred before being stopped by expansion and cooling, no elements heavier than lithium could be formed. (Elements formed during this time were in the plasma state, and did not cool to the state of neutral atoms until much later).

STELLAR NUCLEOSYNTHESIS

Stellar nucleosynthesis occurs in stars during the process of stellar evolution. It is responsible for generation of elements from carbon to calcium by nuclear fusion processes.

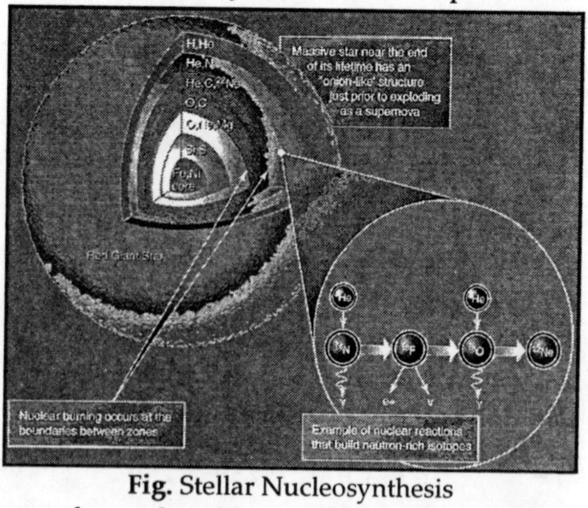

Fig. Stellar Nucleosynthesis

Stars are the nuclear furnaces in which H and He are fused into heavier nuclei, a process which occurs by proton-proton chain in stars cooler than the Sun, and by the CNO cycle in stars hotter than the Sun.Of particular importance is carbon, because its formation from He is a bottleneck in the entire process. Carbon is produced by the triple-alpha process in all stars. Carbon is also the main element used in the production of free neutrons within the stars, giving rise to the s process which involves the slow absorption of neutrons to produce elements heavier than iron and nickel (^{57}Fe and ^{62}Ni). Carbon and other elements formed by this process are also fundamental to life.

The products of stellar nucleosynthesis are generally distributed into the universe as planetary nebulae, or through the solar wind.The first direct proof that nucleosynthesis occurs in stars was the detection of technetium in the atmosphere of a red giant in the early 1950s. Because technetium is radioactive, with halflife much less than the age of the star, its abundance must reflect its creation within that star during its lifetime.

Less dramatic, but equally convincing evidence is of large overabundances of specific stable elements in a stellar atmosphere. An historically important case was observation of barium abundances some 20-50 times greater than in unevolved stars, which is evidence of the operation of the s process within that star. Many modern proofs appear in the isotopic composition of Stardust, solid grains that condensed from the gases of individual stars and which have been extracted from meteorites. Stardust is one component of cosmic dust. The measured isotopic compositions demonstrate many aspects of nucleosynthesis within the stars from which the stardust grains condensed.

EXPLOSIVE NUCLEOSYNTHESIS

This includes supernova nucleosynthesis, and produces the elements heavier than iron by an intense burst of nuclear reactions that typically last but seconds during the explosion of the supernova core. In explosive environments of supernovae, the elements between silicon and nickel are synthesized by fast fusion. Also in supernovae further nucleosynthesis processes can occur, such as the r process, in which the most neutron-rich isotopes of elements heavier than nickel are produced by rapid absorption of free neutrons released during the explosions. It is responsible for our natural cohort of radioactive elements, such as uranium and thorium, as well as the most neutron-rich isotopes of each heavy element.

The rp process involves the rapid absorption of free protons as well as neutrons, but its role is less certain.Explosive nucleosynthesis occurs too rapidly for radioactive decay to increase the number of neutrons, so that many abundant isotopes having equal even numbers of protons and neutrons are synthesized. These include ^{44}Ti, ^{48}Cr, ^{52}Fe, and ^{56}Ni, all of which decay after the explosion to create abundant stable isobars at each atomic weight. Many such decays are accompanied by emission of gamma-ray lines capable of identifying the isotope that has just been created in the explosion.

The most convincing proof of explosive nucleosynthesis in supernovae occurred in 1987 when gamma-ray lines were detected emerging from supernova 1987A. Gamma ray lines identifying ^{56}Co and ^{57}Co, whose radioactive halflives limit their age to about a year, proved that ^{56}Fe and ^{57}Fe were created by radioactive parents. This nuclear astronomy was predicted in 1969 as a way to confirm explosive nucleosynthesis of the elements, and that prediction played an important role in the planning for NASA's successful Compton Gamma-Ray Observatory.

Other proofs of explosive nucleosynthesis are found within the stardust grains that condensed within the interiors of supernovae as they expanded and cooled. Stardust grains are one component of cosmic dust. In particular, radioactive ^{44}Ti was measured to be very abundant within supernova stardust grains at the time they condensed during the supernova expansion, confirming a 1975 prediction for identifying supernova stardust. Other unusual isotopic ratios within those grains reveal specific aspects of explosive nucleosynthesis.

COSMIC RAY SPALLATION

Cosmic ray spallation produces some of the lightest elements present in the universe (though not significant deuterium). Most notably spallation is believed to be responsible for the generation of almost all of ^{3}He and the elements lithium, beryllium and boron (some lithium-7 and beryllium-7 are thought to have been produced in the Big Bang). The spallation process results from the impact of cosmic rays (mostly fast protons) against the interstellar medium.

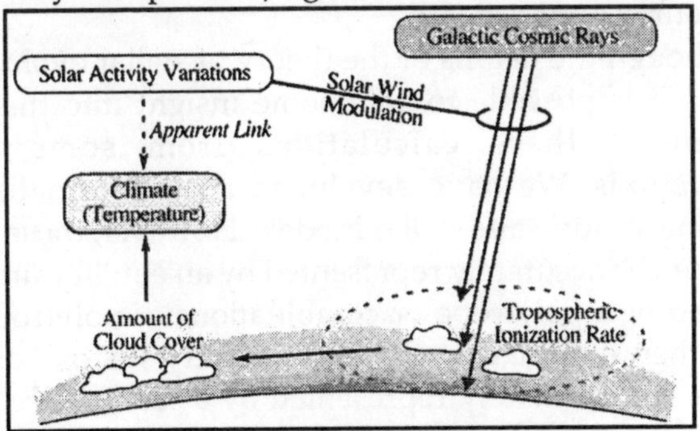

Fig. Cosmic ray Spallation

These impacts fragment carbon, nitrogen and oxygen nuclei present in the cosmic rays, and also these elements being struck by protons in cosmic rays. Beryllium and boron are not significantly produced in stellar fusion processes, because the instability of any ^{8}Be formed from two ^{4}He nuclei prevents simple 2-particle reaction building-up of these elements.

THEORY OF STELLAR EVOLUTION

One of the great triumphs of the twentieth century has been the detailed description of the life history of a star. We now understand with some confidence more than 90 percent of that life history. Problems still exist for the very early phases and the terminal phases of a star's life. These phases are very short, and the problems arise as much from the lack of observational data as from the difficulties encountered in the theoretical description. Nevertheless, continual progress is being made, and it would not be surprising if even these remaining problems are solved by the end of the century.

To avoid vagaries and descriptions which may later prove inaccurate, we concentrate on what is known with some certainty. Thus, we assume that stars can contract out of the interstellar medium, and generally we avoid most of the detailed description of the final, fatal collapse of massive stars. In addition, the fascinating field of the evolution of close binary stars, where the evolution of one member of the system influences the evolution of the other through mass exchange, will be left for another time. The evolution of so-called normal stars is our central concern.

Although the details of the theory of stellar evolution are complex, it is possible to gain some insight into the results expected of these calculations from some simple considerations. We have developed all the formalisms for calculating steady-state stellar models. However, those models could often be accurately represented by an equilibrium model composed of a polytrope or combinations of polytropes. We should then expect that the evolutionary history of a star could be approximately represented by a series of polytropic models. What is needed is to find the physical processes relating one of these models to another thereby generating a sequence. Such a description is no replacement for model calculation for without the details, important aspects of stellar evolution such as lifetimes remain hidden.

In addition, there are branching points in the life history of a star where the path taken depends on results of model calculations so specific that no general considerations will be able to anticipate them. However, a surprising amount of stellar evolution can be understood in terms of sequences of equilibrium models connected by some rather general notions concerning the efficiency of energy transfer. Descriptions of these models, and their relationship to one another, form the outline upon which we can hang the details of the model calculations. In general, we trace the evolution of a star in terms of a model of that star's changing position on the Hertzsprung-Russell diagram. With that in mind, let us briefly review the range of parameters which define the internal structure of a star.

THE RANGES OF STELLAR MASSES, RADII, AND LUMINOSITY

The β^* theorem to show that as the mass of a star increases, the ratio of radiation pressure to total pressure also increases so that by the time one reaches about $100M_\odot$ approximately 80 percent of the pressure will be supplied by the photons themselves. Although it is not obvious, at about this mass the outer layers can no longer remain in stable equilibrium, and the star will begin to shed its mass.

Very few stars with masses above $100M_\odot$ are known to exist, and those that do show instabilities in their outer layers. At the other end of the mass scale, a mass of about 0.1M„ is required to produce core temperatures and densities sufficient to provide a significant amount of energy from nuclear processes. Thus, we can take the range of stellar masses to span roughly 3 powers of 10 with the sun somewhat below the geometric mean.

If we include the white dwarfs, which may be the size of the earth or less, the range of observed stellar radii is about 6 powers of 10 with the sun again near the geometric mean. The observed range of stellar surface temperatures is by far the smallest being from about 2000 K for the coolest M star, to perhaps 50000 K for some O-type stars.

If the temperature range is combined with the range in radii, it is clear that we could expect a range as great as 17 powers of 10 in the luminosity. In practice, the largest stars do not have the highest temperatures, so that the range in luminosity is nearer 10 powers of 10. A reasonable range of these parameters is then

$$10^{-1}M_\odot \le M_* \le 10^2 M_\odot$$

$$10^{-3}R_\odot \le R_* \le 10^3 R_\odot$$

$$10^{-4}L_\odot \le L_* \le 10^6 L_\odot$$

This, then, represents the ranges of the defining parameters of those objects we call normal stars. The theory of stellar evolution will tell us which parameters are related to various aspects of a star's life.

EVOLUTION ONTO THE MAIN SEQUENCE

Problems Concerning the Formation of Stars

The assumption that stars form by contraction from the interstellar medium, honesty requires that we describe several forces that mitigate against that contraction. For a star to form by gravitational contraction from the interstellar medium, all sources of energy which support the initial cloud must be dominated by gravity. For the typical interstellar cloud with sufficient mass to become a star, we shall see that not only is this not true of the collective sources of energy, but also it is not true of them individually.

The Internal Thermal Energy

The internal kinetic energy of the gas of the cloud must be less than one-half the gravitational energy in order for the moment of inertia to show any accelerative contraction. Thus for a uniform density gas at a certain temperature T, the mass must be confined inside a sphere of a certain radius R_c. That radius can be found from

$$2\left(\frac{2\rho\kappa T}{2\mu m_h}\right)\left(\frac{4\pi R_c^3}{3}\right)\leq\frac{GM^2}{R_c}$$

or

$$R_c\leq\frac{GM\mu m_h}{3kT}\approx\frac{0.25(M/M_\odot)}{T}\text{ pc}$$

This distance is sometimes known as the Jeans length, for it is the distance below which a gas cloud becomes gravitationally unstable to small fluctuations in density. For a solar mass of material with a typical interstellar temperature of 50 K, the cloud would have to be smaller than about 5×10^{-3} pc with a mean density of about 10^8 particles per cubic centimeter This is many orders of magnitude greater than that found in the typical interstellar cloud, so it would seem unlikely that such stars should form.

The Rotational Energy

The Virial theorem can also be used to determine the

effects of rotation on a collapsing cloud. Again, the rotational kinetic energy must be less than one-half the gravitational potential energy in order for the cloud to collapse. So

$$2\left(\frac{1}{2}I\omega^2\right) \leq \frac{GM^2}{R_c}$$

which for a sphere of uniform density and constant angular velocity gives

$$R_c \leq \left(\frac{5GM^2}{2\omega^2}\right)^{1/3}$$

The differential rotation of the galaxy implies that there must be a shear or velocity gradient which would impart a certain amount of rotation to any dynamical entity forming from the interstellar medium. For an Oort constant, A = 16 km/s/kpc, this implies that

$$R_c \leq 0.9\left(\frac{M}{M_\odot}\right)^{1/3} \text{ pc}$$

Thus, it would seem that to quell rotation, the initial mass of the sun must have been confined within a sphere of about 0.7 pc.

Magnetic Energy A similar argument concerning the magnetic energy density M, where

$$M = \frac{H^2}{8\pi}\frac{4\pi R_c^3}{3}$$

can be made by appealing to the Virial theorem with the result that

$$R_c \leq 0.37\left(\frac{M}{M_\odot H_{\mu g}}\right)^{1/2} \text{ pc}$$

For a value of the ambient interstellar galactic magnetic field of 5 microgauss we get

$$R_c \leq 0.17\left(\frac{M}{M_\odot}\right)^{1/2} \text{ pc}$$

How are we to reconcile these impediments to gravitational contraction with the fact that stars exits? One can use the rotational and magnetic energies against one another. A moderate magnetic field of a spinning object will cause a great deal of angular momentum per unit mass to be lost by a star through the centripetal acceleration of a stellar wind. The resulting spin-down of the star will weaken the internal sources of the stellar magnetic field itself. Observations of extremely slow rotation among the magnetic A_p seem to suggest that this mechanism actually occurs.

Clouds can be cooled by the formation of dust grains and molecules, as long as the material is shielded from the light of stars by other parts of the cloud. The high densities and low temperatures observed for some molecular clouds imply that this cooling, too, is occurring in the interstellar medium. However, unless some sort of phase transition occurs in the material, the thermal cooling time is so long that it is unlikely that the cloud will remain undisturbed for a sufficient time for the Jeans' condition to be reached. Thus it seems unlikely that the Jeans' condition can be met for low-mass clouds.

It is clear from equation that ñ/T~Rc, so for a given temperature the Jeans' length increases with decreasing density. However, the Jean's mass increases as the cube of the Jeans' length. Thus, for a cloud of typical interstellar density to collapse, it must be of the order of $10^4\ M_{\odot}$. It is thought that the contraction of these large clouds creates the conditions enabling smaller condensations within them to form protostars. The pressure that the large contracting cloud exerts on smaller internal perturbations of greater density may squeeze them down to within the Jeans' length after which these internal condensations unstably contract to form the protostars of moderate mass. These are some of the arguments used to establish the conditions for gravitational contraction upon which all stellar formation depends, and since stars do form, something of this sort must happen.

CONTRACTION OUT OF THE INTERSTELLAR MEDIUM

Since we have given some justification for the assumption that stars will form out of clouds of interstellar matter which have become unstable to gravitational collapse, let us consider the future of such a cloud.

Homologous Collapse

For simplicity, consider the cloud to be spherical and of uniform density. The equation of motion for a unit mass of material somewhere within the cloud is

$$\frac{d^2r}{dt^2} = -\frac{GM(r)}{r^2}$$

If we assume that the material at the centre doesn't move [that is, v(0) = 0], then the first integral of the equations of motion yields

$$v^2 = \frac{8\pi G\rho_0}{3}\int_0^r r dr = \frac{4\pi G\rho_0}{3} r^2$$

or

$$b \propto r$$

This says that at any time the velocity of collapse is proportional to the radial coordinate. This is a self-similar velocity law like the Hubble law for the expansion of the universe, only in reverse. Thus, at any instant the cloud will look similar to the cloud at any other point in time, only smaller and with a higher density $\tilde{n}_0$. Thus, the density will remain constant throughout the cloud but steadily increase with time. Since the velocity is proportional to r, the collapse is homologous and we obtain Lane's law of which completely specifies the internal structure throughout the collapse.

One should not be left with the impression that this homologous collapse is uniform in time. It is not; rather, it proceeds in an accelerative fashion, resulting in a rapid compaction of the cloud. When the density increases to the point that internal collisions between particles produce a pressure sufficient to oppose gravity, the equations of motion become more complicated. Some of the energy produced by the collapse leaks away in the form of radiation from the surface of the cloud and a temperature gradient is established. These processes destroy the self-similar, or homologous, nature of the collapse, and so we must include them in the equations of motion.

However simple and appealing this solution may be, it is a bit of a swindle. The mathematics is correct, and the assumption that v(0) = 0 may be quite reasonable. However, it is unlikely that most clouds are spherically symmetric and of uniform density. Density fluctuations must exist and without the pressures of hydrostatic equilibrium to oppose the central force of gravity, there is no a priori reason to assume spherical symmetry. Normally this would seem like unnecessary quibbling with an otherwise elegant solution. Unfortunately, these perturbations are amplified by the collapse itself and destroy any possibility of the cloud maintaining a uniform density.

Non-Homologous Collapse

Let us consider the same equations of motion as before so that the first integral is given by

$$\int_{v_i^2}^{v^2} dv^2 = 2\int_{r_i}^{r} \frac{4\pi}{3} G\rho_0 r dr = 2\int_{r_i}^{r} \frac{GM(r)}{r^2}$$

Now we wish to follow the history of a point within which the mass is constant so that M(r) is constant and

$$v^2(R) - v_i^2 = 2GM(R)\left(\frac{1}{R} - \frac{1}{R_i}\right)$$

The variable R has replaced r, and this introduces a minus sign into equation since for a collapse dr = -dR. Equation is just the energy integral, so there are no surprises here. Now we change variables so that

$$x = \frac{R}{R_i} \quad \alpha = \frac{2GM(R_i)}{R_i^3} \quad u = \frac{v}{R_i}$$

If we take the initial velocity of the cloud to be zero and the initial value of x to be 1, then equation ecomes

$$u^2 = \left(\frac{dx}{dt}\right)^2 = \frac{\alpha(1-x)}{x} \qquad 0 \le x \le 1$$

which can be integrated over time to give

$$t(x) = \alpha^{-1/2}\left[\frac{\pi}{2} - \sin^{-1}\sqrt{x} + (x - x^2)^{1/2}\right]$$

Now a can be related to the mean density, so that we can rewrite equation as

$$t\left(\frac{R}{R_i}\right) = \left[\frac{3}{8\pi G\overline{\rho_i(R_i)}}\right]^{1/2}$$

$$\left\{\frac{\pi}{2} - \sin^{-1}\left(\frac{R}{R_i}\right)^{1/2} + \left[\left(\frac{R}{R_i}\right) + \left(\frac{R}{R_i}\right)^2\right]^{1/2}\right\}$$

If $\langle \tilde{n}_i(R_i) \rangle$ is constant and not a function of R_i, then we recover the homologous contraction which is clearly not uniform in time. However, if the initial mean density is a decreasing function of R_i, then the collapse time of a sphere of material $M(R_i)$ is an increasing function of R_i. This means that initial concentrations of material will become more concentrated and any inhomogeneities in the density will grow unstably with time.

This is essentially the result found by Larson[1] in 1969. If the cloud is gravitationally confined within a sphere of the Jeans' length, the cloud will experience rapid core collapse until it becomes optically thick. If the outer regions contain dust, they will absorb the radiation produced by the core contraction and reradiate it in the infrared part of the spectrum. After the initial free-fall collapse of a $1M_\odot$ cloud, the inner core will be about 5 AU surrounded by an outer envelope about 20000 AU When the core temperature reaches about 2000 K, the H_2 molecules dissociate, thereby absorbing a significant amount of the internal energy.

The loss of this energy initiates a second core collapse of about 10 percent of the mass with the remainder following as a "heavy rain". After a time, sufficient matter has rained out of the cloud, and the cloud becomes relatively transparent to radiation and falls freely to the surface, producing a fully convective star. While this scenario seems relatively secure for low mass stars (i.e., around $1M_\odot$), difficulties are encountered with the more massive stars. Opacities in the range of 1500 to 3000 K make the evolutionary tracks somewhat uncertain. Indeed, there are some indications that massive stars follow a more homologous and orderly contraction to the state where they become fully convective.

Although this is the prevailing picture for the early phases of the evolution for low-mass protostars, there are some difficulties with it. Such stars would be shielded from observation by the in-falling rain of material until quite late in their formation. Since the entire configuration including the rain is hardly in a state of hydrostatic equilibrium, the arguments given below would not pertain until quite late in the star's formation, by which time the star may well have reached the main sequence. There seems to be little support in observation for this point of view, and the entire subject is still somewhat controversial.

Michael Disney[2] has pointed out that the details of the collapse from the interstellar medium depend critically on the ratio of the sound travel time to the free-fall time in the contracting protostar. Although this ratio is typically unity small departures from unity appear to matter. The free-fall time is basically the time during which the collapse takes place, and the sound travel time is the time required for the interior to sense the effects of pressure disturbances initiated at the boundary. Thus, if $\tau_s/\tau_f > 1$, the interior tends to be unaffected by the boundary pressure during the collapse. Any external pressure will then tend to compress the matter in the outer part of the collapsing cloud without affecting the interior regions, removing any density gradients that may exist in the perturbation and forcing the collapse to be more nearly homologous.

This would reduce the effect of the rain and cause the protostar to collapse more as a unit. Any initial velocity resulting from of the homologous collapse of the large cloud will only exacerbate the situation by significantly shortening the time required for the collapse.

Thus the initial phases of star formation remain in some doubt and probably depend critically on the circumstances surrounding the initial conditions of the collapse of the larger cloud.

CONTRACTION ONTO THE MAIN SEQUENCE

Once the protostar has become opaque to radiation, the energy liberated by the gravitational collapse of the cloud cannot escape to interstellar space. The collapse will slow down dramatically and the future contraction will be limited by the star's ability to transport and radiate the energy away into space. Initially, it was thought that such stars would be in radiative equilibrium and that the future of the star would be dictated by the process of radiative diffusion in the central regions of the star.

Indeed, for most stars this is true for the phases just prior to nuclear ignition. However, Hayashi[3] showed that there would be a period after the central regions became opaque to radiation during which the star would be in convective equilibrium.

Hayashi Evolutionary Tracks

At transporting energy as long as there are no sources of energy other than gravitation, the future contraction will be limited by the star's ability to radiate energy into space rather than by its ability to transport energy to the surface. We have also learned that the structure of a fully convective star will essentially be that of a polytrope of index n = 1.5. We may combine these two properties of the star to approximately trace the path it must take on the Hertzsprung-Russell diagram.

With gravitation as the only source of energy and the contraction taking place on a time scale much longer than the dynamical time, the Virial theorem allows one-half of the change in gravitational energy to appear as the luminosity and be radiated away into space. The other half will go into the internal energy of the star increasing the internal temperature. Thus,

$$L = \frac{1}{2}\left(\frac{d(GM^2/R)}{dt}\right) = -\frac{\frac{1}{2}GM^2}{R^2}\frac{dR}{dt}$$

Since the luminosity is positive, dR/dt must be negative which ensures that the star will contract. Since the luminosity is related to the surface parameters by

$$L = 4\pi R_*^2 \sigma T_e^4$$

the change in the luminosity with respect to the radius will be

$$\frac{dL}{dR_*} = \frac{4L}{T_e}\frac{dT_e}{dR_*} + \frac{2L}{R_*}$$

Equation is essentially a definition of what we mean by the effective temperature. As long as the star remains in convective equilibrium, it will be a polytrope and the contraction will be a self-similar, and thus homologous, contraction.

Since the rate of stellar collapse is dictated by the photosphere's ability to radiate energy, we should expect the photospheric conditions to dictate the details of the collapse. Indeed, as we shall see in the last half of this book, the eigenvalues that determine the structure of a stellar atmosphere are the surface gravity and the effective temperature.

So as long as the stellar luminosity is determined solely by the change in gravity, and the energy loss is dictated by the atmosphere, we might expect that the independent variable T_e to remain unchanged. However, it is necessary to show that such a sequence of models actually forms an evolutionary sequence. The extent to which this will be true depends on the radiative efficiency of the photosphere. This is largely determined by the opacity.

At low temperatures the opacity will increase rapidly with temperature owing to the ionization of hydrogen. This implies that any homological increase of the polytropic boundary temperature at the base of the atmosphere will be met by an increase in the radiative opacity and a steepening of the resultant radiative gradient. This increase in the radiative opacity also forces the radiating surface farther away from the inner boundary, causing the effective temperature to remain unchanged.

The star can effectively be viewed as a polytrope wrapped in a radiative blanket, with the changing size of the polytrope being dictated by the leakage through the blanket. The blanket is endowed with a positive feedback mechanism through its radiative opacity, so that the effective temperature remains essentially constant. The validity of this argument rests on the ability of convection to deliver the energy generated by the gravitational contraction efficiently to the photosphere to be radiated away. With this assumption, we should expect the effective temperature to remain very nearly constant as the star contracts.

THE STRUCTURE AND EVOLUTION OF MAIN SEQUENCE STARS

When the track of a gravitationally contracting star intersects the main sequence, the star has reached a point in its life when it will be stable for an extended period of time. This is ensured observationally by the fact that about 90 percent of all stars reside on or very near the main sequence and so must be involved in the utilization of their most prolific and efficient source of energy " the fusion of hydrogen into helium. So we may take the intersection of the Henyey track with the main sequence as an indication that hydrogen ignition has begun in the stellar core.

Actually nuclear processes begin somewhat before the main sequence is encountered. The first constituents of the star to undergo nuclear fusion are deuterium and lithium which require conditions substantially below that of hydrogen for their ignition. However, their abundance is sufficiently low so that they provide little more than a stabilizing effect on the star as it proceeds along its Henyey track, causing the star to hook on to the main sequence.

For the sun, about a million years is required for the equilibrium abundances of the proton-proton cycle to be established with sufficient accuracy for their use in the energy generation schemes. At this point, the star can be said to have arrived at the zero-Age main sequence. As the name implies, this is generally taken as the beginning point of stellar evolution calculations, as the onset of nuclear burning makes the details of the prior evolution largely irrelevant to the subsequent evolution.

In some real sense, the star forgets where it came from. Since it is fairly obvious that the effects of stellar evolution during the main sequence phase will result in little movement on the H-R diagram, we need to understand more of the structure of the interior to appreciate these effects. Therefore, we begin by describing the structure to be expected for the hydrogen burning models that describe the main sequence. The structure of main sequence stars can be readily broken into two distinct groups: those that occupy the upper half of the main sequence, and those that occupy the lower main sequence.

LOWER MAIN SEQUENCE STARS

We define the lower main sequence to be those stars with masses less than about 2 solar masses. For these stars, after the trace elements with low ignition temperatures have been exhausted and hydrogen fusion has begun, the equilibrium structure is established in about a million years. The mass of these stars is insufficient to produce a central temperature high enough to initiate the CNO cycle, so the primary source of energy is the proton-proton cycle. Models indicate that in the sun, 98 percent of the energy is supplied by the proton-proton cycle. The relatively low dependence on temperature of the proton-proton cycle implies that the energy generation will be less concentrated toward the centre than would be the case with the CNO cycle. This and the modest central temperature imply that a temperature gradient less than the adiabatic gradient is all that is required to carry the energy produced by the p-p cycle. Thus, these stars have a central core which is in radiative equilibrium.

However, in the sun, the adiabatic gradient is never far from the actual temperature gradient, and it would take a very little increase in the temperature gradient to cause the core to become unstable to convection.

Indeed, the conditions for convective instability are met in the outer regions of these stars resulting in the formation of a convective envelope. In the sun, this point is reached at about $0.75R_{TM''}$, so that about 98.8 percent of the mass is included in the radiative core. Ultimately, the situation is reversed near the surface, as it must be, for the energy leaves the surface of the star by radiation.

The existence of the radiative core in stars of the lower main sequence has a significant effect on the subsequent evolution of the star. The ^{4}He, which is the end product of hydrogen burning, remains in the locale in which it is produced. However, since the production rate is strongly dependent on temperature, the helium abundance increases more rapidly as one approaches the centre of the star. The helium must be supported against its own gravity while it contributes nothing to the support of the remainder of the star.

As a result, the internal temperature will increase to maintain the luminosity in the face of decreasing hydrogen abundance and the increasing mass of the particles (i.e. the ^{4}He). This is why the temperature scales with the mean molecular weight m. Thus, we should expect stars like the sun to slowly increase in brightness, as the internal temperature rises, during their main sequence lifetime. Indeed, the standard solar model indicates that the solar luminosity has increased by about 40 percent since its arrival on the zero age main sequence.

Toward the end of the star's main sequence life, the helium abundance will rise to the point where a core of helium, surrounded by a hydrogen burning shell, will form in the centre of the star. The support of this isothermal helium core is eventually helped by the Pauli Exclusion Principle. We outlined the equation of state to be expected for a gas where all the available h^3 volumes of phase space were filled. Because of their lower mass, this condition will be experienced first by the electrons.

The degenerate equation of state does not contain the temperature and therefore permits the existence of an electron pressure capable of assisting in the support of the helium core; this equation is independent of the conditions existing in the hydrogen-burning shell. Thus as the core builds, we could expect its structure to shift from that of an isothermal sphere, to that of a polytrope with a ã= 5/3, as would be dictated by the electron pressure of a fully degenerate gas. This change from an isothermal sphere to a polytrope will dictate the mass distribution, for the pressure of the ions becomes small compared to that of the electrons. However, because of the high conductivity of a degenerate gas, the configuration will remain isothermal since any energy surplus can immediately be transported to a region of energy deficit by electron conduction. Thus, the region is still known as the isothermal helium core, even though the pressure equilibrium is dictated by the electron pressure behaving as a polytropic gas with a ã of 5/3.

Therefore, the main sequence lifetime of a low mass star consists of a steady energy output from hydrogen burning in an environment of steadily increasing helium. On a nuclear time scale, the helium abundance increases preferentially in the most central regions causing the temperature to rise which results in a slow increase in the luminosity throughout the main sequence lifetime of the star. After about 10 percent of the radiative core mass has been consumed, an isothermal helium core begins to form and structural changes begin to occur very rapidly. This signals the end of the main sequence lifetime.

Upper Main Sequence Stars

The situation regarding the stellar structure for stars of more than 2 solar masses is nearly reversed from that of the lower main sequence. For stars on the main sequence, the observed mass-radius relation is approximately

$$M \propto R^{4/3}$$

Therefore, for stars along the main sequence, we expect the central temperature to increase slowly as we proceed up the main sequence in accord with

$$T_c \propto R^{1/3} \propto M^{1/4}$$

This slow rise in the central temperature will result in a greater fraction of the energy being produced by the more temperature-sensitive CNO cycle. Thus, by the time one reaches stars of greater than about 2 solar masses, the CNO cycle will be the dominant source of energy production. The much larger temperature sensitivity of the CNO cycle as compared to the p-p cycle means that the region of energy production will be rather more centrally concentrated than in stars of less mass.

This, in turn, requires a steeper temperature gradient in order to transport the energy to the outer parts of the star. Since in the sun the radiative gradient was already quite close to the adiabatic gradient, this small increase is sufficient to cause the inner regions to become convectively unstable, and a substantial convective core will be established. However, in the outer parts of the star, the declining density causes the product of êñ, which appears in the radiative gradient to reduce the radiative gradient below that of the adiabatic gradient, and so convection stops. Thus, we have a star composed of a convective core surrounded by an envelope in radiative equilibrium. This role reversal for the core and envelope has a profound effect on the evolution of the star.

The presence of a convective core ensures that the inner regions of the star will be well mixed. As helium is produced from the burning of hydrogen, it is mixed thoroughly throughout the entire core. Thus, we do not have a buildup of a helium core that increases in helium abundance toward the centre in these stars. Instead, the entire convective core is available as a fuel source for energy production at the centre of the star. For this reason, energy production is remarkably steady in these stars until the entire convective core is nearly exhausted of hydrogen.

Even as exhaustion approaches, the extreme temperature dependence of the CNO cycle implies that deficits produced by the declining availability of hydrogen fuel can be made up by modest increases in the temperature and hence minor changes in the structure of the star. Indeed, it is not until more than 99 percent of the convective core mass has been converted to helium that truly significant changes occur in the structure of the star and the star can be said to be leaving the main sequence.

POST MAIN SEQUENCE EVOLUTION

The evolution of stars off the main sequence represents the response of the star to a depletion of the available fuel supply, and it can be qualitatively understood by examining the response of the core and envelope to the attempts of the nuclear burning regions to adjust to the diminution of the available hydrogen.

In stars of the lower main sequence, the hydrogen-burning shell begins to move into a region of declining density resulting in a decrease in available hydrogen. For stars of the upper main sequence, the situation is somewhat different. The convective nature of the core ensures the existence of mass motions, which continue to bring hydrogen into the central regions for hydrogen burning until the entire core is depleted. These two rather different approaches to hydrogen exhaustion produce somewhat different evolutionary futures for the two kinds of stars, so we examine them separately.

Evolution off the Lower Main Sequence

The development of a helium core, which signals the onset of post main sequence evolution, is surrounded by a thin hydrogen-burning shell. The hydrogen burning continues in a shell around the helium core which steadily grows outward, in mass, through the star.

However, the helium core must be supported against its own gravity as well as support the weight of the remaining star, and its energy sources are all on the outside. As a result, it is impossible for the hydrogen-burning shell to establish a temperature gradient within the helium core. Only gravitational contraction of the core will result in the release of energy inside the helium core, and except for this source of energy the helium core must be isothermal, with its temperature set by the burning of hydrogen surrounding it. But the rate of hydrogen burning is dictated largely by the mass of material lying above the burning zone, because this is the material that must kept in equilibrium. As the mass of the isothermal helium core increases, the equilibrium temperature of the core will also rise and this demand can be met only by a slow contraction of the helium core. The slow increase in the core temperature triggers a steady increase in the stellar luminosity.

As the isothermal core grows through the addition of He from the hydrogen-burning shell, the core temperature must rise in order for it to remain in equilibrium and support the outer layers of the star. Since an isothermal sphere is a unique polytropic configuration, it seems reasonable that there would be a limit to the amount of overlying material that such an isothermal core could support. This limit is known as the Chandrasekhar-Schönberg (C-S) limit. The limit will depend solely on the mass fraction of the isothermal core and the mean molecular weights of the core and envelope. Should the core exceed this limiting mass fraction, it must contract to provide the temperature and pressure gradients necessary to support the remainder of the star as well as itself.

Degenerate Core

Only for stars near the upper end of our range will the mass of the core approach the Chandrasekhar-Schönberg limit without becoming degenerate and the core undergoing further gravitational contraction. For stars with masses M d" 1.3M„ the slowly developing isothermal core will be degenerate from a point in its development when the core mass is well below the Chandrasekhar-Schönberg limit. Under these conditions, that limit does not apply because the added pressure of the degenerate electron gas is sufficient to support nearly any additional mass. Thus, the isothermal helium cores of lower main sequence stars can increase to virtually any mass below the Chandrasekhar degeneracy limit. As mass is added to the core. This law, following differentiation with respect to time, indicates that the core will shrink on the same time scale that mass is added to it, and that is the nuclear time scale.

Progress to the Red Giant Phase

In terms of its physical size, this isothermal degenerate helium core is never very large. Thus the post main sequence evolution of a low-mass star can be viewed as the processing of stellar material through the burning zone, with the resultant helium being packed into a very small volume of systematically higher mean molecular weight. The declining density just above the helium core will lead to an increase in temperature, in order for the nuclear energy generation mechanisms to supply the energy required to support the star. However, an increase in the central temperature would lead to an increase in the temperature gradient and an increase in the luminosity.

The increased luminosity, in turn, causes the outer envelope of the star to expand, decreasing the temperature gradient. Equilibrium is established at a higher shell temperature and somewhat greater luminosity and temperature gradient. The result is that the star moves upward and very slightly to the right on the H-R diagram. The process continues until the temperature gradient exceeds the adiabatic gradient. Then the entire outer envelope becomes convective. The increase in physical size of the envelope lowers the surface temperature and thereby increases the radiative opacity in the outer layers. This further decreases the efficiency of radiative transport and hastens the formation of the outer convection zone. The outer envelope is now well approximated by a polytrope of index n = 1.5, and the conditions for the Hayashi tracks become operative.

The star now approximately follows the track of a fully convective star only now in reverse. The continual decline of the available hydrogen supply in the shell burning region, which becomes extremely thin, leads to a steady increase in the shell temperature and accompanying rise in the luminosity. With the outer convection zone behaving as a good polytrope and efficiently carrying the energy to the surface, the energy loss is again limited by the photosphere and the star expands rapidly to accommodate the increased energy flow. The star now moves nearly vertically up the giant branch as a red giant.

Helium Flash

As the temperature of the hydrogen-burning shell increases and the degenerate core builds in mass; the temperature eventually reaches approximately 10^8 K. This is about the ignition temperature of helium via the triple-± processes. Under normal conditions, the burning of helium could begin in a measured way which would allow for an orderly transition of nuclear energy generation processes.

However, the core is degenerate, so the electron pressure is only weakly dependent on temperature. Indeed, the limiting equation of state for total degeneracy does not contain the temperature at all.

Thus helium burning sets in with unrelenting ferocity. With its extreme dependence on temperature, the triple-± process initiates a thermal runaway which is limited only by the eventual removal of the degeneracy from the core. The complete equation of state for a partially degenerate gas does indeed, contain the temperature and at a sufficiently high temperature the equation of state will revert to the ideal-gas law. When this occurs, the core rapidly expands, cools, and reaches equilibrium, with helium continuing to burn to carbon in its centre.

The response of the core to this entire process is so swift that the total energy produced is a small fraction of the stored energy of the star. In addition, the site for the production of the energy is sufficiently far removed from the outer boundary that energy is diffused smoothly throughout the star and never makes a noticeable change in the stars appearance.

The duration of the flash, is so much shorter time than the dynamical time scale for the entire star that one would expect that all manifestation of the flash would be damped out by the overlying star and remain hidden from the observer. However, detailed hydrodynamical calculations[7] indicate that the pressure pulse resulting from the rapid expansion of the core arrives at the surface with a velocity well in excess of the escape velocity. This may well result in a one-time mass loss of the order of 30 percent which would affect the subsequent evolution.

TERMINAL PHASES OF LOW MASS EVOLUTION

Initially, after helium ignition, the hydrogen burning shell continues to supply about 90 percent of the required support energy. However, now an orderly transition of energy mechanisms can take place, resulting in the transfer from hydrogen burning to helium burning over an extended time. The star will move somewhat down the giant branch and out on the horizontal branch, from near the peak of the giant branch where the helium flash took place. Meanwhile the helium core is in convective equilibrium, with the convection zone extending almost to the hydrogen shell.

The re-expansion of the core is responsible for the contraction of the outer envelope, causing the star to move out onto the horizontal branch. It appears likely, that after helium burning has ceased and the resultant carbon core is contracting, the outer envelope becomes unstable to radiation pressure and lifts off the star, forming a planetary nebula and leaving the hot core, which now relieved of its outer burden, simply cools. If the mass is below the Chandrasekhar limiting mass for carbon white dwarfs, the star continues to cool to the virtually immortal state of a white dwarf.

STRUCTURE AND EVOLUTION OF WHITE DWARFS

Abundant stellar component of the galaxy derived the equation of state appropriate for a relativistic and a nonrelativistic degenerate gas and found them to be polytropic. The mass-radius relation for polytropes in general, which provides the approximate results appropriate for white dwarfs. There are basically two approaches to the theory of white dwarfs. The first is to observe that a relativistically degenerate gas will behave as a polytrope and to explore the implications of that result. The second is to investigate the detailed physics that specifies the equation of state and to create models based on the results. Cox and Giuli[8] and references therein provide an excellent example of the latter. Our approach will be much nearer the former.

The ejection of a planetary nebula during the later phases of the evolution of a low-mass star leaves a hot degenerate core of carbon and oxygen exposed to the interstellar medium. While such a core may range in mass from about 0.1 of a solar mass to more than a solar mass, its future will be remarkably independent of its mass. While the actual run of the state variables will pass through regions of degeneracy through partial degeneracy to a nondegenerate surface layer, the basic properties of the star can be understood by treating the stars as polytropes. From observation we know that the white dwarf remains of stellar evolution are about a solar mass confined to a volume of planetary dimensions and thus will have a density of the order of $\rho = 10^6$ g/cm^3. If we assume that the gas is fully ionized, then the typical energy of an electron will be about 0.1 MeV for a fully degenerate gas.

If the stellar core were at a temperature of 10^7EK, the typical ion would have an energy of about 1 keV. Since energy densities are like pressures, even if the number densities were the same for the electrons and ions, the pressure of the electrons would dominate. In fact, since the typical ion produces many electrons, the dominance of the electron pressure, is even greater.

Thus the structure will be largely determined by the electron pressure and the ions may be largely ignored. However, Hamada and Salpeter[9] have shown that at densities around 10^8 g/cm^3 the Fermi energy of the electron "sea" becomes so high that inverse beta decay becomes likely and some of the electrons disappear into the protons of the nuclei, causing the limiting mass to be somewhat reduced over what would be expected for a purely degenerate gas. Also, the thermal energy of the ions is lost, permitting the star to shine.

Since these stars are largely degenerate, most of the momentum states in phase space are full, and an electron that is perturbed from its place in phase space has to travel quite a distance before it can find an empty place.

This implies that the mean free path of electrons will be very long in spite of the high densities. Such electrons play essentially the same role as the conduction electrons of a conductor so that the electrical and thermal conductivity will be very high in a degenerate gas. This crowding of the states in phase space also results in the reduction in radiative opacity since it is difficult for a photon to move an electron from one state to another. As a result, it is very difficult for temperature gradients to exist within a fully degenerate configuration. However, in the outer regions of the white dwarf where the gas becomes partially degenerate, the opacity rapidly rises and the conductivity drops, giving rise to a steep temperature gradient with the result that the energy flow to the surface is seriously impeded. Thus Aller has likened a white dwarf to a metal ball wrapped in an insulating blanket. Since the structure of a polytrope is stable and independent of the temperature, the evolutionary history of a white dwarf largely revolves on the details of its cooling.

In 1952, Leon Mestel[10] took basically this classical approach to the cooling of white dwarfs and found that the cooling curve [dln(L)/dln(t)] was approximately constant and independent of time. Iben and Tutukov, using a much more detailed analysis and equation of state, found virtually the same result which they regarded as occurring through a series of accidents. Their results give [dln(L)/dln(t) H" (-1.4, -1.6)] for $5 < \log(t_{yrs.}) < 9.4$. It is true that a considerable number of effects complicate the simple picture of a polytrope wrapped in a blanket.

For example, while we may neglect the ions for an excellent approximation of the description of the white dwarf structure, the contribution of the ion pressure will make the star slightly larger than one would expect from the polytropic approximation. Because of the extreme concentration of the star, a small contraction produces a considerable release of gravitational energy, which is then to be radiated away. This extends the cooling time significantly over that which would be expected simply for a cooling polytrope. Toward the end of the cooling curve a series of odd things happen to the equation of state for the white dwarf. As the interior regions cool, they undergo a series of phase transitions first to a liquid state and then to a crystalline phase.

Each of these transitions results in a "heat of liquefaction or crystallization" being released and increasing the luminosity temporarily. Problems of the final cooling remain in the understanding of the low-temperature high-density opacities that will determine the flow of radiation in this final descent of the white dwarf to a cool cinder, called a black dwarf in thermal equilibrium with the ambient radiation field of the galaxy. The question is of considerable interest since such objects could be detected only by their gravitational effect and could bear on the question of the "missing mass".

EVOLUTION AWAY FROM THE UPPER MAIN SEQUENCE

The evolution of the more massive stars that inhabit the upper main sequence is driven by the same processes that govern the evolution of the lower main sequence, namely, the exhaustion of hydrogen fuel. However, the processes are quite different. The exhaustion of the convective core leads to the production of a helium centre, as in the lower main sequence stars, but now the core will have to contend with the Chandrasekhar-Schönberg limit.

Nature of the Massive Helium Core

In massive stars, as the hydrogen is depleted in the helium core, the temperature rises rapidly, to produce the energy necessary to accommodate the demands of stellar structure. For stars with masses greater than about 7 $M_\odot$, the resulting helium core will be greater than the Chandrasekhar-Schönberg limit; and to make up for the energy deficit caused by the failing hydrogen burning, the core will have to contract. Since the contraction must maintain a temperature gradient, the contraction will proceed rather faster than would be expected for an isothermal core. However, the steep temperature gradient established by the terminal phases of core hydrogen burning will be relaxed because the energy generated by gravitational contraction will not be as centrally concentrated as it was from hydrogen burning. This drop in temperature gradient will cause convection to cease, yielding a core in radiative equilibrium supplying the required stellar energy by contraction.

The contraction of virtually any polytrope will result in an increase in internal temperature. This is really a consequence of the Virial theorem. However, the cessation of hydrogen burning in the core and the resultant decrease in the temperature gradient imply a change in the overall structure of the core, and thus the polytropic analogy is somewhat strained. The decline in the temperature gradient actually implies that the core will suffer a reduction of its internal energy while the boundary temperature increases.

This loss of internal energy goes into the support of the outer envelope. Thus, both this energy and the energy generated by contraction are available for the support of the outer layers of the star. The increase in the boundary temperature of the core will result in a slow expansion and cooling of the radiative envelope for the same reason described for low-mass stars. After a suitable rise in the boundary temperature, hydrogen is reignited in a shell surrounding the helium core.

This results in a marked change in the temperature gradient of the hydrogen-burning shell at the core boundary. The localization of the energy production in such a small region steepens the temperature gradient to the point where the temperature gradient exceeds the adiabatic gradient, driving the entire envelope into convection. The outer envelope now rapidly transfers the energy to the surface, which again becomes the limiting barrier to its escape. The star moves rapidly toward the giant branch as a star with a helium core surrounded by a hydrogen-burning shell and covered with a deep convective envelope. This envelope is so deep that it reaches into the region where nuclear processing has taken place, dredging up some of this material to the surface. The result of this process is evident in the atmospheric spectra of some late-type supergiants.

Stars with masses less than about 5M„ will end their hydrogen burning with a helium core below that of the Chandrasekhar-Schönberg limit and will contend with a slowly contracting isothermal core right through the ignition of a hydrogen-burning shell. The future of such a star is mirrored in the behaviour of lower-mass stars except that helium ignition takes place before the core becomes significantly degenerate. The result is that these stars experience no helium flash, and the transition to helium burning is orderly.

Ignition of the Massive Helium Core

In both cases described above, the contraction of the helium core proceeds while the hydrogen shell is burning. In the more massive stars, where the core is above the

Chandrasekhar-Schönberg limit, the star must do so to maintain a temperature gradient for its own support. In the less massive case, the core grows slowly as a result of the processing of the hydrogen in the energy-generating shell. The added mass results in a slow core contraction. In both instances, the decreasing density in the hydrogen-burning shell necessitates a rise in the temperature required for energy generation.

After the increasing temperature gradient caused by this increasing shell temperature has forced the envelope to become fully convective, the convective envelope continues to expand, for the energy escape is again limited by the radiative efficiency of the photosphere. Eventually the central temperature reaches 10^8 K, and helium ignition takes place. The ignition has a dramatic effect on the core but does not exhibit the explosive nature of the helium flash. The core undergoes a rapid expansion and, because of the huge temperature dependence of the triple-a process, becomes unstable to convection.

This produces an expanded convective helium core surrounded by a hydrogen-burning shell. This shell supplies more than 90 percent of the energy required to maintain the luminosity. The rapid core expansion is accompanied by a contraction of the outer envelope with a corresponding increase in the surface temperature. The star moves off to the left in the H-R diagram, maintaining about the same total luminosity. The hydrogen burning shell continues to supply the majority of the energy throughout the helium burning phase which proceeds in much the same manner as the main sequence core hydrogen-burning phase, but on a much shorter time scale.

TERMINAL PHASES OF EVOLUTION OF MASSIVE STARS

The end phase of the evolution of massive stars is still somewhat murky and the subject of active research. Initially, the helium burning continues in the core until the core becomes largely carbon. At the point of helium exhaustion, the core again gravitationally contracts, rising in temperature, until helium is ignited in a shell source around the core. For stars with a mass between 3 and 7 solar masses, there is some evidence that the carbon core which develops is degenerate and that ignition, when it occurs, occurs explosively, perhaps producing a supernova.

For stars of more than 10 solar masses, carbon burning can take place in a nonviolent manner, producing cores of neon, oxygen, and finally silicon, each surrounded by a shell source of the previous core material which continues to provide some energy to the star. The results of silicon burning yield elements of the iron group for which further nuclear reactions are endothermic, and so this burning will not only fail to contribute energy for the support of the star but also rob it of energy. In addition, the densities become high enough that the electrons are forced into the protons of the nuclei by means of inverse b decay. This effect has been called neutronization of the core. Both mechanisms produce a significant number of neutrinos which also do not take part in the support of the star against gravity and can be viewed as "cooling" mechanisms for the core.

This rapid cooling precipitates a rapid collapse of the core followed by the entire star. The in-fall velocity soon exceeds the speed of sound, resulting in the formation of a shock wave, and interior densities may become large enough for the material to become opaque to neutrinos. Under some conditions the endothermic nuclear reactions may bring about the disintegration of the iron-group elements into ^4He. Which process dominates for stars of which mass is not at all clear. The shock wave formed by the in-fall may "bounce", or the increased neutrino opacity may provide sufficient energy and momentum to ensure that a large fraction of the star will be blown into the interstellar medium. The remnants, if any, could be a neutron star or a black hole. Although the details of the terminal phases of massive stars remain somewhat unclear, it is virtually certain that these phases are likely to end with the production of a supernova and the subsequent enrichment of the interstellar medium by heavy elements.

Chapter 7

The Mountains

GEOSYNCLINES

The early Paleozoic rocks of the Interior Lowland of North America are everywhere only a few hundred or at most a few thousand feet thick. But in the Appalachian Mountains, rocks of the same age are many times this thickness. Furthermore, the rocks of the Interior Lowlands are largely limestone and dolomite, with subordinate clastics, whereas clastic rocks or metamorphic rocks derived from them predominate in the Appalachians. Thus the rocks of the mountain belt are both much thicker and more generally clastic than those of the same age in the continental plate to the northwest.

Geologists have since found comparable facies changes associated with almost all the other ranges: the Alps, Himalayas, Urals, Andes, and the Cordilleran System of North America. In many of these, thicknesses of as much as 6 or even 8 miles of strata have been measured, whereas the average thickness over the adjacent continental plates is not more than a few thousand feet. The strata comprising all of these great chains are chiefly volcanics and sediments or their metamorphic derivatives.

Some preserve rainprints, mudcracks, salt-crystal casts, and other indications that they were deposited in shallow water, and some, like the coals of Pennsylvania, are even land-laid beds. Fossils found in them are nearly all of shallow-water organisms. Though a few strata in the Alps, West Indies, and Timor have been interpreted as deep-sea oozes, their associations make this doubtful, and, in any case, their volume is trivial. These great piles of strata do not, then, represent the filling of all ocean deep, even though they are thick enough to fill any existing deep to overflowing, for shall-water deposits recur time after time through sections as thick as 40,000 feet. Only one conclusion seems possible: The crust in these areas was being depressed at a rate roughly (equal to that of sediment accumulation; the surface of the sediments was never far above or below sea level during all time of their accumulation.

The basement beneath these accumulations must have warped slowly downward; the trough thus formed has been called a *geosyncline,* and the pile of strata filling it a *geosynclinal prism*. The association Hall recognized, of mountain chains with geosynclines, has since been found generally, though not quite universally. Mountains thus differ from continental plates in a fundamental way:

THEIR SEDIMENTARY THICKNESSES ARE FAR GREATER

The thickening and coarsening of sediments from the stable continental plate into the Appalachians suggest that most of them were derived from a land southeast of the present mountains—either the Piedmont, the Coastal Plain, or, perhaps, an area now covered by the Atlantic. Similar relations have been found for other geosynelines.

Inferences as to the exact area furnishing the sediment cannot be certain, however, within wide limits, as a glance at the map of Burma will show. The Irrawaddy dumps its huge load 450 miles from the delta of the Brahmaputra—a distance as great as that from Pittsburgh, Pennsylvania, to Birmingham, Alabama—yet both rivers derive much of their load from within a few miles of each other in the eastern Himalayas. Many modern examples of this sort show how hazardous is precise inference as to the source of ancient sediments, even though the point where they entered the ancient sea can be quite closely determined.

Though the exact sources of sediments are uncertain, there is less ambiguity about the general position of the shoreline from which they entered the seas. Thus three elements have been recognized in common association: the *geosyncline* itself, with its thick pile of strata; the *foreland*, a part of the continental plate covered only by relatively thin sediments; and, on the opposite side of the geosyncline, the *hinterland*, which supplied the bulk of the sediments but of which we know little. Obviously, strata equivalent to those of the geosyncline must be lacking or sparse in the hinterland, for this was undergoing erosion. Also, many of the postulated hinterlands are now veneered with younger rocks or drowned beneath the sea.

Though the association of geosynclines with most mountain ranges is unmistakable, the picture given in the preceding paragraph is too broad to apply to all ranges. For example, the Yellow Sea has many features suggestive of a geosyncline. This shallow sea is the resting place for the great load of detritus brought from western China by the swollen torrents of the Yellow River and the Yangtze.

The deltas of these streams are like the ancient Devonian delta that has been recognized in the rocks of the Catskills and Appalachians of New York and Pennsylvania, and which was built by streams entering the Appalachian geosyncline from the vanished hinterland to the cast. But the "foreland" of the Yellow Sea is deep sea fringed by islands, not part of the stable continental plate of Asia. Similarly, the accumulation of Cretaceous, Tertiary, and Quaternary sediments along the coast of the Gulf of Mexico is clearly approaching geosynclinal dimensions, but the "foreland" of the geosynclinal prism is the deep water of the Gulf.

These modern analogs of the ancient geosynclines do not fit simply into a pattern like the Paleozoic Appalachian geosynclines or the Mesozoic and Tertiary Alpine geosyncline. Yet, the similarities in abrupt thickening of strata to prodigious amounts along a fairly simple belt warrant calling them geosynclinal. Quite possibly, in the distant geologic future a great mountain range will rise along the now swampy coast of Louisiana and Mississippi, but present knowledge does not fully warrant such a prophecy. For example, a thick series of pre-Cambrian strata in western Montana, although now a part of the towering Rockies, lay essentially undisturbed in their depositional attitude through all of Paleozoic and most of Mesozoic time—roughly 400 million years—despite the fact that they are more than 30,000 feet thick and cover an area half as large as the Alps. Also, considerable mountain masses, such as Ruwenzori in Africa, the highlands of Labrador, and the Adirondack Mountains, are not built from deformed geosynclinal sediments.

In summary, we can say that the association of mountain ranges with geosynclinal prisms is far too common to be accidental—there must be some,genetic connection between them—but the quiescence of geosynclinal accumulations as thick as those of most mountain ranges through hundreds of millions of years, and the occurrence of mountains not made of thick elastic sediments, show that geosynclines are not essential prerequisites of mountain uplift. Factors other than excessively thick sediment accumulation are needed for mountain-making.

Mountain Structure

We have mentioned that most mountains differ from the continental plates in their much more deformed rocks. A study of the structural details of some of the most completely investigated ranges permits some inferences as to the forces that deformed them.

THE APPALACHIANS

The Appalachians rise from beneath the coastal plain of Alabama and extend northeastward in a sinuous belt 1,500 miles long to Nova Scotia and New Brunswick, where their deformed rocks are cut off by the Atlantic shore. In general, individual faults and folds lie about parallel to the trend of the belt and persist for many miles, so that geologic cross sections spaced several miles apart generally resemble their neighbors closely.

In the central Appalachians the folds toward the northwest are open, but, as one proceeds southeastward, the folds become tighter, with steeper limbs. The northwesterly of these tighter folds are merely asymmetrical, with the northwest limbs steeper than those on the southeast, but farther southeast the folds are overturned, so that both limps dip southeast. Locally, at the southeast side, thrust faults have repeated certain beds of Cambrian age as many as five times. Furthermore, at the northwest the rocks are unmetamorphosed, but as they are traced southeastward they become more altered, and the southeast side of the mountain belt is underlain by slates, schists, and gneisses, with some unmetamorphosed granite plutons.

In West Virginia and to the south, the main mass of the folded belt is locally thrust several miles northwestward over the rocks of the foreland. These great thrusts, some of which have been folded, were first recognized from the pattern of the outcrops; their existence has been conclusively confirmed by oil wells drilled through rocks of the upper plate into younger strata beneath the fault.

In the latitude of Knoxville as many as six large thrust faults have been recognized. Great thrust faults also mark the northwest border of the folded belt in the Hudson-Champlain Valley in northern New York and Vermont. Some of these faults have rocks of the same age both above and below them. But the rocks of the two plates are of wholly different facies: One is largely carbonate rocks and the other is metamorphosed elastic rocks. In other words, rocks of the same age, but deposited so far apart that their composition is wholly different, have been brought in contact by thrusting.

Crustal Shortening

A series of symmetrical anticlines and synclines might be pictured as forming in either of two ways: by differential uplift and sinking of the basement on which the folded strata lie or, perhaps, by a force which pushes the edges of the strata horizontally so that they wrinkle and warp like a pile of blankets on a table when shoved together. But systematically asymmetric folds that all lean to the northwest and thrust faults that have brought rocks northwestward for many miles as we see them in the Appalachians cannot be explained by vertical movements. Here the rocks must have been shoved together horizontally. When the folds of the leastdeformed part of the Appalachians—in Pennsylvania—are straightened out in imagination, by measuring the length of beds following all the convolutions shown in geologic sections and then comparing this with the width of the folded belt, it has been found that points originally 81 miles apart have been brought nearer together by 15 miles. In other words, the superficial crust of the earth has locally been shortened 18 per cert.

This measurement takes no account of the shortening by thrust faults farther to the southeast. In the southern Appalachians, where half a dozen thrust faults in a single cross section each required several miles of overriding, the shortening must have been far greater. Similar studies show comparable or greater shortening in many other ranges. As an example we will cite the best-studied range on earth, the Alps.

THE ALPS

The Alps lie in a great are looped northward through Switzerland to Vienna, where they die out beneath the Hungarian plains. They are part of the great mountain system that extends from the Spanish Pyrenees across southern Europe and Asia through the Himalayas and the Malay Peninsula.

Like the Appalachians, the Alps rise on the site of a great geosyncline, but the strata that filled this trough are of Mesozoic and early Cenozoic age rather than Paleozoic. The foreland lies to the north, where relatively thin Mesozoic rocks rest unconformably on Paleozoic and older rocks. Within the Alps Mesozoic and Tertiary rocks of wholly different facies and far greater thicknesses are found.

Despite the marvelous cliff exposures, Alpine structure is so complex and the facies changes are so great that the arduous labour of hundreds of geologists during more than a century of intensive study still leaves many problems unsolved, but the broader features are well established. A much simplified map of the Alps.

Divisions of the Alps

As in the Appalachians, the structure and history of the

different segments along the trend of the range differ. The major division is between the western and eastern Alps, for there is a profound difference both in facies and structure along an irregular line trending south from Lake Constance along the Rhine just west of the Silvretta Alps. The western Alps lie chiefly in Switzerland, though their western are extends into France and the southern parts are in Italy. The eastern Alps are chiefly in Austria, with their northern border in Bavaria and their southern in Italy.

The Jura Mountains

The Jura Mountains consist of foreland rocks of Mesozoic age folded into anticlines separated by almost flat synclines. The amplitude of the anticlines increases southward, and some of the southern ones are broken by small thrust faults. Significantly, even the largest anticlines expose no rocks older than Middle Triassic. This fact, together with exposures in railroad tunnels and in areas north and west of the mountains, suggests that the Jura folds were formed by the crumpling of a thin sheet of sedimentary rocks that has been torn loose from its foundation and pushed northward.

The geometry of the folds does not permit them to extend downward into the Lower Triassic, Permian, and older rocks that are known to underlie the mountains. Clearly, the surficial rocks were sheared from the nearly flat beds beneath, much as we can crumple a tablecloth by sliding it over the table. This is merely the outermost and least of many Alpine examples of large-scale horizontal movement of the rocks, yet the surface rocks in the southern Jura have moved several miles from the site of their deposition.

The Swiss Plateau

Between the Jura and the Alps lies the Swiss plateau with the beautiful lakes Geneva, Neuchatel, Zurich, and Constance. Most of the rocks are Tertiary, chiefly coarse conglomerates derived from the Alps to the south. This is shown both by their southward coarsening and by the pebbles, which can commonly be matched with their parent formations, though, as we shall see, not always with their present neighbors. Though broadly synclinal, this area is locally strongly folded, and along its southern border is overriden by great thrust sheets.

The Prealps

Between the Swiss plateau and the main limestone Alps is a chain of lower but still impressive mountains that range from about 3,500 to 6,500 feet above sea level. In places they are separated from the main range to the south only by a zone of saddles in their high projecting spurs. These are the Prealps. Their magnificent cliffs expose chiefly Mesozoic rocks that rest in thrust contact on the crumpled Tertiary of the Swiss plateau.

Thus they are thrust sheets whose rocks are completely foreign to the basement on which they lie. Such thrust masses that have been carved by erosion from a formerly continuous sheet are called *klippes*. Similar masses have been recognized in many other mountain chains.

Portrays the kind of evidence used in interpreting the complex geology of the Prealps. A brief study of this map shows that the klippes of the Mythen and Rotenfluh are merely the highest of at least three thrust sheets piled one on top of the other. At the south end of Lake Lucerne the Eocene rocks clearly dip beneath Jurassic rocks which are overlain by Cretaceous and Eocene. These rocks pass at Sisikon and Bauen beneath the Cretaceous and Eocene on which rest the Triassic of the Rotenfluh and Mythen.

THE HIGH LIMESTONE ALPS AND THE CRYSTALLINE MASSIFS

The lower thrust sheets beneath the klippes of the Prealps continue into the High Limestone Alps. They are largely the sheared-off limbs of great anticlinal folds that have been overturned so far that they lie nearly flat. Such sheets are called by the Swiss geologists *nappes* or *Decken*. South of them, where the plunge of the folds allows them to be seen, lie the highly sheared anticlinal cores from which the sedimentary mantles have been nearly completely severed and shoved northward to make the nappes of the Limestone Alps. These cores consist of granite, gneiss, schist, and other metamorphic rocks such as are seen in the magnificent tower of Mont Blanc. A few highly distorted plant fossils have been found in anthracite coal associated with these metamorphic rocks; from them we know that at least part of the core is of Carboniferous age. The overlying sedimentary rocks have yielded some Permian and Triassic fossils and contain pebbles of the core metamorphics.

We thus have in the cores of the folds from which the nappes of the High Limestone Alps have been sheared off and thrust northward a rock succession much like that of the foreland north of the Jura.

In both places, Permian and Triassic rocks rest with great unconformity on steeply tilted and locally highly metamorphosed Carboniferous rocks associated with pre-Permian intrusive masses. The relation of these "crystalline (metamorphosed) massifs" to the mantles that have been stripped from them to form the High Limestone Alps, and of these rocks, in turn, to the thrust sheets of the Prealps, can be seen in the map and sections much simplified from the detailed original mapping of the great Swiss geologist Argand.

The Swiss geologists consider the Mont Blanc massif and others like it to have moved only a few kilometers. By contrast, the great thrust sheet of the High Limestone Alps has been torn loose and overfolded northward like cloth slid over a table and crumpled into a flat fold. This sheet can be followed foot by foot for more than ten miles to the northwest, where it disappears beneath the higher thrust sheet of the Prealps. This is evidence of considerable horizontal movement in the earth's crust. What, then, are we to think when, several miles farther northwest, identical rocks reappear from beneath the Prealpine sheet and rest in fault contact on the Tertiary of the Swiss plateau?

One strongly overfolded crystalline core after another is found in the southern Alps, each separated from the other by highly deformed schists. Some of these schists have yielded Mesozoic or early Tertiary fossils. The geometrical arrangement of the great thrust masses demands tremendous shortening of the visible crust—a shortening that cannot, of course, be accurately measured, but which certainly must amount to many scores of miles. Thus the crust beneath northern Italy must lie scores, if not, indeed, a hundred or more miles, closer to the Swiss plateau than it did in Eocene time!

The evidence is not only geometric. The facies of the rocks composing the thrust masses demand great travel also. For example, Arnold Heim mapped the Cretaceous of the nappes of the High Limestone Alps in eastern Switzerland. Their pattern justified his drawing the cross section reproduced, much simplified, at the top of the diagram. The distortion of the rocks along both faults suggested that the overlying rocks moved northward with respect to those beneath, on both thrusts. The facies of the Cretaceous rocks in the nappes practically demand this type of movement.

For example, if one wished to regard the upper thrust sheet (now exposed to the north of the lower) as having always been north of it, and, similarly, that the Cretaceous covering the Aar massif has always been south of the lower nappe as it now is, one must accept the unlikely arrangement of sedimentary facies shown in the second diagram. If, however, we overrule our "common sense" reaction, that such great travel is "unreasonable," and accept the geometrical evidence at its face value, we can reconstruct the logical arrangement of sedimentary facies shown in the lowest diagram. Obviously, this facies arrangement is not only consistent with the geometrical evidence, but it offers almost conclusive proof that both lower and upper nappes traveled northward, and that the upper sheet has come from farther south than the lower.

Summary of Alpine Structure

Although the few examples cited can only hint at the almost incredibly complex structure of the Alps, they suffice to show that prodigious horizontal forces must have operated here, just as they did in the Appalachians. The crust has been shoved together as though by the jaws of a great vise, and the material caught between the jaws has been squeezed out and shoved northward relative to the underlying rocks for distances measured in tens of miles. In the process, the rocks have been dynamically metamorphosed—all transitions can be found in a particular rock mass between almost unaltered shale and highly metamorphosed schist. Granitic rocks caught in the deformation have been partly or wholly converted into gneiss, with all transitional stages apparent. These are significant observations, for both geometrical patterns and metamorphic rocks identical with those now found in the towering Alps are repeated again and again in the great pre-Cambrian shield areas of the earth.

Time of Deformation

The Alpine folding took place in many stages, but most of it was in Oligocene and early Miocene time, as rocks of these ages are involved in the thrusting, whereas the Upper Miocene and Pliocene rocks nearby are much less deformed. The Appalachian folding occurred through a long period, as recorded by angular unconformities within the Ordovician of New York and Quebec, between Ordovician and late Silurian in the Hudson Valley, within the Devonian of the Canadian Maritime Provinces, and, greatest of all, between Carboniferous and Triassic. Abrupt coarsening of sediments at several other times during this long period suggests other pulses of folding that have not been proved by unconformities.

The folds and faults of the Appalachians as we now see them may be said to date chiefly from the late Paleozoic. In Scotland and Scandinavia all these features can be matched in mountains whose deformation dates from Silurian times. Examples could be given from many other parts of the world. It is tempting to conclude that the linear ranges were all formed by the crumpling and piling together of geosynclinal sediments by horizontal compressive forces operating at fairly shallow levels in the earth's crust. Certainly, most ranges do show features that demand crustal shortening. But it would be premature to assert that all do.

THE BASIN RANGES

Most of Nevada, western Utah, and large parts of Oregon,

Idaho, Arizona, New Mexico, west Texas, and eastern California constitute a geologic province characterized by isolated mountain ranges separated by desert plains and basins. Some of these ranges are relatively low, but others exceed 12,000 feet in altitude and 100 miles in length. Indeed, the westernmost of these ranges, the Sierra Nevada, is about the size of the Alps.

As Dutton long ago said, on a relief map of North America the ranges resemble an army of caterpillars crawling toward Mexico. The geology of this large area is far less well known than that of the Appalachians or the Alps; only small segments have been mapped in detail.

Yet our information shows that these mountains differ notably in structure and history. One fact is very impressive: These mountains were formed from parts of the crust that in themselves show widely differing earlier histories, and not from a uniform geosyncline.

Much of the *Basin Range* province was the site of an early Paleozoic geosyncline; Cambrian and Ordovician rocks attain thicknesses of many thousands of feet in eastern California and southern Nevada.

In later Paleozoic time, however, the region was occupied by two smaller subsiding basins separated by a highland, for there are many thousands of feet of Devonian and Carboniferous rocks in western Utah and eastern Nevada, and again in western Nevada, whereas in parts of central Nevada the Permian rests directly on Ordovician.

Whether a true mountain range was formed at this time, and its possible extent if it were, are questions that await further detailed mapping. This point is almost irrelevant to our present purpose, however, for whatever the size and trend of the emergent area, the Mesozoic history of the different parts of the province was very different. Western Nevada had a huge geosynclinal accumulation of Triassic rocks; the Triassic is thin and continental in southern Nevada and absent from much of western Utah and southern Arizona.

Similarly, the Jurassic is geosynclinal in western Nevada and continental in southern Nevada. Cretaceous rocks are unknown in western Nevada and over much of western Utah; they are thin and continental in southern Nevada, but form a thick geosynclinal accumulation in southern Arizona and New Mexico.

The Nevada Jurassic geosyncline underwent strong folding in early Jurassic and again in late Jurassic time. Following the second of these episodes the great plutons of the Sierra Nevada were emplaced in the folded belt. Long erosion uncovered these plutonic rocks, and they were afterward flooded by Tertiary volcanic fields.

As a result of these various histories, the ranges differ greatly in structure and internal composition, although their gross forms and relations to the lowlands are closely similar. For example, the Sierra Nevada (largely plutonic, with local patches of Tertiary volcanics on its crest and western slope), Steens Mountain, Oregon (almost wholly of Tertiary lava flows), the Wasatch Range (pre-Cambrian, Paleozoic, and Mesozoic folded rocks cut by early Tertiary plutons), and the Little Hatchet Mountains of southern New Mexico (chiefly Lower Cretaceous sedimentary rocks with some Tertiary volcanics) have had such different early histories that we recognize significant variations from the folded ranges.

Basin Range Structure

What is the feature common to the Basin Ranges? Though not all the ranges have been mapped even in reconnaissance surveys, it appears significant that all those in which alluvium does not mask the critical features give evidence of uplift along normal faults.

The observed faults dip at angles ranging between 40° and the vertical, averaging nearly 60°. Where the body of the range consists of folded rocks or of thrust masses, these normal faults cut across the older folds and faults. Where the intervening basins are not masked by alluvium it can be shown that, as in Pleasant Valley west of the Sonoma Range, they are downdropped with respect to the ranges.In other words, the Basin Ranges, unlike those described above, owe their existence directly to differential vertical movements of adjacent blocks in the earth's crust. No lateral compression of the surface rocks is necessary.

MOUNTAIN STRUCTURE

Let us recapitulate the features in which mountains differ from the continental plates:

- The sial beneath the mountains is far thicker than that beneath the continental plates. This generalization is supported by gravity measurements in many areas, by seismic data for many others. In the Colorado Rockies it is less directly shown by a heat flow sufficiently larger than average to suggest the existence there of an unusually thick prism of highly radioactive, and hence sialic, rocks. We have insufficient data to support such a statement with regard to the Basin Ranges as a whole, and in view of the narrowness of many of these ranges, it seems improbable that the blocks composing them contain sial notably thicker than that beneath the adjacent basins.
- The rocks of the mountains are more deformed than those of the continental plates. This generalization seems clearly valid for nearly all ranges, though again we must note that as far as the Basin Ranges are concerned it does not apply to those composed largely of Tertiary volcanic rocks. Nor does it apply in an obvious cause-and-effect relation to the nonvolcanic ranges of this province, for the complex structures they display are far older than the uplift of the present ranges and have not been intensified during uplift.
- Granitic plutons are confined to mountains (and to shield areas). This general statement seems valid, though plutons younger than the folding are only of trivial extent in the Alps and are absent from many ranges such as the California Coast Ranges.
- Sedimentary thicknesses are greater in the mountain belts than in the continental plates. This generalization

is well supported by data from many ranges. But it does not apply as a distinguishing feature of the mountains and basins of the Basin Ranges. Nor, as was noted in connection with the pre-Cambrian of Montana, does unusual sedimentary thickness suffice to cause mountain-making. And it fails to apply on a local scale, for it is extremely doubtful whether the Colorado Rockies ever had eroded from them Cretaceous deposits as thick as those found in the nearby intermont basins, which locally contain as much as 20,000 feet of rocks of this age.

- In many of the great mountain chains the rock structures are such as to demand large horizontal shortening of the surficial crust. Again, the Basin Ranges offer a conspicuous exception.

What is the meaning of these generalizations in terms of mountain origin? We find no specific features that we can identify as the "jaws" of a vise that has squeezed together wide segments of the crust into a narrow belt. How has this come about? We will return to this enigma after we consider the subsequent history of such a compressed belt, however it was formed.

THE SUBSEQUENT HISTORY OF FOLDED BELTS

Whatever the mechanism by which tightly folded belts are made, isostasy demands that the average level of the base of the folded sedimentary prism must sink. Local masses may be sustained by the strength of the crust, but excess loads of regional extent cannot be. The sial composing the outer crust has a density of about 2.7. We do not know the precise density of the underlying sima, but it is certainly greater or isostasy could not prevail. Estimates from seismic data range between 3.0 and 3.3 for the density just below the Mohorovièiæ discontinuity. Let us take the higher of these figures. If, by shortening of a belt of the outer crust, it is thickened 1 km., in order to maintain isostatic balance the basement must sink far enough to displace a mass of sima equal to that of the added sial.

This means that the bottom would sink $\frac{2.7}{3.3} \times 1km.$, or more than 800 meters. Thus if we imagine an overthrust sheet 1 km. thick thrust over a wide expanse of country, the surface elevation would increase only about 200 meters. Such an estimate as this cannot be taken literally, as the quantitative values to be assigned are too indefinitely known.

But there can be little doubt of the general validity of the reasoning, for it agrees with two significant facts about folded mountains: the long persistence of uplift along them and their association with granitic plutons. It also conforms with some leading facts regarding the geology of the great oceanic deeps.

Several features testify to some uplift and concurrent erosion of mountain ranges during their folding. Among these are: unconformities within the geosynclinal prism, the general presence of coarse clastic deposits in the outer gently folded parts of mountain chains, and the common presence in them of pebbles that could have come only from rocks of far greater age exposed by deep erosion.

Apparently, however, much of the uplift of all ranges takes place by vertical bowing upward, after folding has ceased. Perhaps the most compelling evidence of this are the unconformities. At their southwest end, the Appalachians fall off in altitude and disappear as a topographic unit beneath the Cretaceous beds of the Gulf Coast. The map of the surface relief formed during the folding of the southern Appalachians had been entirely erased by erosion before the Cretaceous deposits were laid down. Yet the present Appalachians attain their greatest height at Mount Mitchell, about 200 miles along the strike to the northeast.

Less convincing but still strong evidence of long-delayed uplift of many ranges is found in the erosion surfaces carved across them. The profiles of present streams have characteristically high gradients in their source areas. In many mountain ranges, however, the streams have their sources in summit uplands of relatively low relief, whose gentle sloping surfaces cut smoothly across rocks that vary greatly in composition and structure, without reflecting these differences in hill and valley forms. The area must once have been a low plain on which streams of low gradient were working. Yet these uplands are now dissected by deep trenches and canyons, with the foaming torrents of modern streams tumbling down them. The contrast in gradient strongly suggests that the summit upland was formed when the streams all had low gradients such as could have existed only when the area stood much nearer to sea level than at present. The evidence is very strong that the mountain mass had been nearly or almost erased as a topographic feature before it was again uplifted.

When we analyse this situation, our idealized picture of the thickened prism of sial beneath the mountain belt may suggest part of the answer. If the range is to remain in isostatic balance, far more than 1 km. of material must be eroded from the surface in order to lower the summit of the range 1 km. The situation is the reverse of that prevailing at the time of thrusting. In fact, the erosion of 1 km. of material would reduce the average elevation by only about 200 meters under the simplified assumptions of our illustration. And since erosion by streams with concave side-slopes removes a very large proportion of the mass before the divides are greatly lowered, it is entirely possible that the elevation of the divides will actually increase through isostatic uplift at the same time that the average elevation of the block is decreasing by valley cutting.

THE MOUNTAIN ROOT

So far we have considered the thickened sial as maintaining all its properties unchanged, but this is certainly not true. All rocks whose structural relations show evidence of deep burial and deformation are now metamorphosed. Clearly, this requires high temperatures. Mere burial beneath other rocks will raise the temperature of the geosynclinal sediments by blanketing in the earth's internal heat; in addition, the rocks contain radioactive elements whose disintegration contributes to the heating. If a thick column of sialic rock is heated, it may expand enough to raise the surface very notably. Inasmuch as heating is slow, a long time may elapse before the surface of the prism rises because of this effect. Perhaps this is an important factor in the long-delayed uplift of many ranges.

But if the burial is to very great depths, such as the 60-km. root of the Sierra Nevada, heat flow is extremely slow, and bowed-down sial may come in close contact with hot sima. Under these conditions it may gradually heat up to the point of fusion, forming a mass of granitic magma. The abundance of migmatites in deeply eroded mountain chains suggests such refusion, and so, too, does the practical confinement of granitic plutons to mountain chains. But a root so softened by partial melting can no longer maintain itself as a root; it must gradually spread sideways and flatten out. In so doing, it would thicken the sial alongside the range, perhaps accounting in this way for the high-standing plateaus or high plains that border many ranges, for example the Rocky Mountains and Himalayas.

THE SHIELD AREAS

Each of the continents contains large areas over which pre-Cambrian rocks are at the surface, among them most of eastern Canada, Finland and Sweden, northeastern Siberia, India, much of South Africa, western Australia, and eastern Brazil. They are called the *pre-Cambrian shields*. At their borders most of them are overlain by Cambrian rocks, resting upon an unconformity of relatively low relief.

Most of the shield areas are only a few hundred feet above sea level, and the character of the surrounding sediments of younger age suggests that this has been their condition through most of later time. However, not all the shield areas stand low: the Adirondacks of New York (an outlier of the Canadian shield), much of Labrador, the mountainous areas near Lake Baikal on the Siberian shield, and large parts of the South African and Brazilian shields stand a few thousand feet above the sea.

They are as high, for example, as much of the Urals and Appalachians (which were folded near the end of the

Paleozoic), or the Scottish Highlands (where the folding was near the middle of the Paleozoic). Yet these are exceptional parts of the shields, most of them lie close to sea level.

Owing to their lack of fossils, it is impossible to correlate pre-Cambrian rocks over wide areas. Accordingly, in the shield areas we cannot reconstruct the geology with anything like as much confidence as for such ranges as the Appalachians. Nevertheless, despite this handicap, the structures in much of the shield can be deciphered well enough to prove the presence of old mountain chains.

The evidence lies in the structural patterns which are brought out by detailed mapping. The similarity in complexity and general patterns is apparent. On the average, the shield areas have considerably more granite. Furthermore, migmatites, which are rare or absent from granitic borders in most of the younger ranges), are abundant. They make up about a quarter of the exposed rocks in Finland. It seems reasonable to infer that in these belts we have the deeply eroded roots of ancient mountains.

The cores of many Paleozoic and younger ranges contain more intensely deformed rocks than their external zones, and the extent of the metamorphic rocks is proportionately greater in older ranges than in mid-Tertiary ranges like the Alps. Such metamorphic cores are absent from most late Tertiary and Quaternary ranges. The contorted and metamorphosed zones of the pre-Cambrian shields thus seem what we should expect at lower and lower erosion levels in a folded mountain chain.

Perhaps the generally low topography and structural stability of the shield areas—for it must be remembered that the shields project below the flatlying strata of the continental plates and thus underlie most of the continents at a depth averaging less than a mile—are due to long-continued erosion of their surfaces, combined with lateral spreading of the former mountain roots until, in the long course of geologic time, the sial beneath them is just thick enough to maintain them in isostatic balance near sea level.

Further erosion is very slow. They may be further depressed by sedimentation, but for the most part they are stable. Why some relatively small parts of them may, after long standing in this low position, rise to mountainous heights, as the Adirondacks, is a problem for which no generally satisfactory answer has yet been found.

OCEAN DEEPS AND THE BELTS OF NEGATIVE ANOMALIES

The many earthquakes that occur close to the deeps imply crustal unrest in these areas, as does the common, though not invariable, association of active volcanoes with the island arcs. As the deeps are the maximum departures of the earth's surface from sea level, and many of them immediately adjoin large islands or even the continent of South America (Atacama Deep, Leeward Trough), the deeps must be fairly young geologically or they would have been filled with sediment. We have noticed that the sediment dredged from the Philippine Deep east of Mindanao is obviously derived from that island, and had both deep and island existed for a geologically long time, the deep would surely have been partly filled, if not destroyed.

One of the greatest geological discoveries of the last generation was that there are remarkable deviations from isostasy associated with some of the island arcs and ocean deeps. This discovery we owe to the distinguished Dutch geodesist F. A. Vening-Meinesz, who, between 1923 and 1932, determined the value of gravity at many points at sea.

The earth's surface were level and the crust homogeneous, the attraction of gravity would be determined entirely by the latitude at which the observition is made. The anomalies shown on this chart were determined by subtracting this theoretical value of gravity from the actual value, after certain corrections have been made.

The corrections include allowances for the depth of the submarine beneath the surface, the topography of the land and sea floor for long distances around the station, the fact that sea water differs greatly in density from the rocks underlying land stations, and, finally, an allowance for "regional compensation" (equivalent to assuming that a very large crustal block centering at the station is in isostatic balance with the rest of the crust). These actual computations are laborious and far beyond the scope of this book, but the meaning of them is not. The great belt of negative anomalies can mean only that the belt is underlain by rocks whose density is far less than is usual at equivalent depths elsewhere—in other words, by light rocks. The belt is so narrow—only 30 to 100 miles in the strongly anomalous part—that the deficiency in gravity demands a very deep septum (30-40 miles) of light rocks to account for it.

The belt is about 3,000 miles long. For much of its course it lies close to the axis of a deep oceanic trough, lie on either side, and others overlie a submarine ridge that divides the deep in two. In Timor, Tanimber, Kei, and Ceram Islands, this ridge comes to the surface, but the negative strip continues. On both sides of the negative strip the gravity anomalies become positive over broad, not obviously linear areas. The positive anomalies are considered to mean that the denser rocks deep in the crust are closer to the surface than they normally are, and hence exert greater than normal attraction.

To summarize, along the negative strip light crustal rocks are greatly thickened and extend to unusual depths, forming a long narrow wedge in the denser rocks on either side.

Now these very large deviations from isostatic equilibrium are not at all common. The uplift of Scandinavia after removal of the glacial ice indicates that even in the shield areas (long unaffected by intense crustal movements or igneous activity since before the Paleozoic, and which might, therefore, be considered unusually strong) the crust has responded to far smaller forces than these great density differences indicate in the East Indies.

Both the numerous earthquakes and widespread active volcanoes indicate that the crust is probably unusually weak along the East Indian are, as compared, for example, with one of the shield areas. Vening-Meinesz pointed out that these facts can only be reconciled by assuming that the rocks of the negative strip are being *currently held down so that they are prevented from floating at the level appropriate to their density*. In other words, some selective downward force of very great magnitude is now operating here.

The rocks within these belts must be under very great shearing stress, as they are being dragged down against their natural tendency to float upward. Clearly, such conditions might produce folding and thrusting of the type we associate with mountain chains. Conforming with this deduction, the islands along the belt are, indeed, composed of highly deformed rocks are locally capped by warped and uplifted coral reefs. Do we have here a mountain chain in process of formation?

SPECULATIONS REGARDING MOUNTAIN-MAKING

Although the student will recognize that the mere statement of a geologic "fact" almost of necessity carries with it an element of inference, the facts so far stated are generally regarded by geologists as well established. However, the objectives of science include not merely the description of natural phenomena, but also the coordination of data into broader generalizations, and the fitting of these into a comprehensive theory. Yet, though most geologists agree on the facts and even on most of the generalizations hardly a one would state that any satisfactory theory has been advanced to explain the making of mountains. Yet theories and speculations are valuable in the progress of science, for they suggest tests to be made which may modify or confirm some of the basic generalizations. What are some of the speculations suggested to explain the great enigma of mountain-making?

Theory of a Shrinking Earth

One speculation favored by many geologists in the early days of geology (and supported today by some) was that the earth is a shrinking body. In shrinking, the more rigid rocks of the crust must accommodate themselves to a smaller inner core by crumpling, just as the skin of an apple crumples and wrinkles when it dries. It was assumed that the deformation of rocks along mountain chains has been localized by the unusual thickness of relatively weak sediments in the geosynclines. The earth is losing heat at the surface, as is proved by the increase of temperature measured in mines and oil wells. Before radioactivity was discovered, it seemed certain, therefore, that the earth must be cooling off and, hence, contracting.

Although many geological objections have been made to this theory, perhaps the principal one is that shrinking should shorten every great circle equally. The present distribution of folded mountains is far from that to be expected on this theory. A second argument is less easily tested, but seems valid: The Alpine-Himalayan chain is essentially the only considerable eastwest chain on earth.

Considering scale and strength, it seems unlikely that the crust is strong enough to transmit the great forces half-way around the earth so as to concentrate all the shortening into this single belt. Also, the distances between fold belts of other trends seem far too great. The most serious objection, however, and one that most geologists regard as insuperable, is that, from reasonable estimates of the distribution of radioactive materials in the earth, it seems probable that all the heat now escaping from the surface is of radioactive origin. The earth as a whole is probably not becoming appreciably cooler. Indeed, it may be heating up, for computations of heat transfer show that even in 2 billion years the earth cannot have lost any appreciable heat by conduction from beneath a depth of 700 km. Until these objections are overcome, this once popular theory seems discredited.

THEORY OF CONTINENTAL DRIFT

Probably every schoolboy who has studied a globe has independently discovered that if the Americas were pushed eastward they would almost fit the coast of Africa and Europe. This rough "fit" is even more striking when we notice that the Mid-Atlantic Ridge swings sharply eastward at about the right place to fit the Guinea Coast of Africa. Nearly a century ago, Antonio Snider published the speculation that the continents east and west of the Atlantic had formerly been joined and have since drifted apart.

This idea seemed so fantastic to geologists of that time that it was lost sight of until about 40 years ago, when it was again independently suggested by the American geologist F. B. Taylor and the German Alfred Wegener. If whole continents can move so far, the shortening of the chest by a few scores of miles in the mountain belts becomes insignificant. Any force that could move the Americas westward two or three thousand miles could furnish the Rockies and Andes as a minor by-product. Although a mechanism adequate for bodily displacing the continents is unknown, this theory has, nevertheless, won favour from many geologists.

Wegener assumed that the continents float about on the denser substratum almost like rafts on a millpond, and that the ocean floors and the deep layers beneath are so weak that they yield almost like a fluid to very small forces. Two forces tend to move the continents. The first is centrifugal: The sial masses float higher than the sima and so are further from the earth's axis. Hence the centrifugal force acting on them is greater, and they tend to drift toward the equator. The second is the tidal attraction of the sun and moon which, as the earth rotates from west to east, tends always to pull the continents westward. Wegener thought the Alpine-Himalayan chain was formed when Eurasia collided with Africa and India in response to the first force, and that the Andes and Rockies were piled up by friction as the Americas were dragged westward by the tidal pull.

Both these forces exist, but computations show that they are millions of times too small to have folded the rocks. If so, one may wonder why the theory is still seriously considered. But many geological phenomena are well established—continental glaciation, for one—that have so far defied explanation. If geologists were convinced that the continents were formerly more closely grouped, they would accept the fact, even though no adequate mechanism has yet been suggested. Let us briefly consider the geological evidence advanced in support of such a grouping.

Obviously, more than the apparent "fit" of the Atlantic shores is needed. Indeed, this "fit" at a depth of 10,000 feet is far less impressive, and there were notable differences in the continental outlines in Cretaceous or even in mid-Tertiary time. The main arguments advanced are the similarities in the trends of mountain belts, in strata, and in fossils of land-dwelling plants on opposite sides of the Atlantic.

In New England and Newfoundland, the fold axes of mountain belts formed during the middle and late Paleozoic are cut off by the shore. Mountain structures of similar, but not identical, ages in the British Isles and western Europe also end at the Atlantic shore. Likewise, preDevonian fold axes in Brazil and Uruguay end at the Atlantic, as do some folds of about the same age in South Africa. The apparent interruption of these fold lines by the ocean suggests, but does not prove, that they were formerly joined and have since drifted apart. The hazard in such an assumption is shown by the fact that the mid-Tertiary chains of the Atlas and Pyrenees ranges also end at the shore, and some geologists have tried to find their western continuation in the Greater Antilles.

It is clear from the pattern of the gravity anomalies and island arcs, however, that the Antilles structures never extended eastward to join them, but curve southward through the nearly drowned are of the lesser Antilles to join the Venezuelan ranges. It is as likely that the mid-Paleozoic folds of Newfoundland curved slightly to connect with those of east Greenland as that they joined those of the British Isles.

More suggestive, though still open to question, are similarities in the strata of South America and Africa. In both regions the oldest fossil-bearing rocks are Upper Silurian or Devonian. The fossils resemble those of the Falkland Islands but differ from those of northern lands. In both regions similar continental rocks overlie these marine beds, and both areas contain late Carboniferous tillites.

Tillite of about the same age is found also in South Australia, Madagascar, and India. Many fossil plants are associated with this tillite, and coal beds overlie it. In all these widely separated areas two genera of fernlike plants, *Gangamopteris* and *Glossopteris*, occur, but they are found in the northern hemisphere only in Russia, northeastern Siberia, and India. The *Glossopteris* flora has been cited as "proving" a complete separation of the southern lands and India from the rest of the northern hemisphere, and the name "Gondwana Land" has been given this hypothetical continent that included all the areas with these plants. However, such vagaries in plant distribution are not rare. Many Permian plants of Texas and Arizona are not known to occur in the more easterly parts of North America, nor in Europe, but are found in China. These are never cited as evidence of a former union between Asia and North America.

Perhaps the strongest argument for a drastically different pattern of continents in late Paleozoic time lies in the glacial pattern. Glaciers invaded South America from the east and the Falkland Islands from the north, apparently radiating from a point now in the South Atlantic. Glaciers entered Australia from the south as far as latitude 28° S. An ice cap centered near 15° S in Africa, leaving its northern moraines on the equator. The Indian glacier centered in the Aravalli Mountains, north of Bombay, in latitude 20° N. Characteristic rocks from here are found in the tillite hundreds of miles away, both to the southeast and northwest. How are we to account for continental ice sheets in the present tropics?

One suggestion is that the earth's poles have shifted in position, and another, sometimes linked with it, that the now-scattered parts of Gondwana Land were formerly a single mass near the South Pole.

Mere shifting of the earth's axis without moving the land masses would not avoid the embarrassment of continental ice sheets in the tropics, for the tillites are now so placed that no possible shift would fail to leave one or another within 20* of the equator. Wegener thought, therefore, that in the late Paleozoic the Gondwana fragments were gathered into a single mass, with the South Pole at about the common junction of Africa, Australia, Antarctica, India, and Madagascar. He thought the American and Europian coals represented tropical peat bogs, whereas the *Glossopteris* coals were formed in peat bogs like those of Ireland or Alaska.

The tillite near Boston, within a few miles of the Rhode

Island coals, is thus interpreted as that of a mountain glacier. But the Russian and Siberian Glossopteris localities must have been in the tropics.

Others besides Wegener have tried to explain the climates of the late Paleozoic by different groupings of the continents. All require special explanations for nearly as many phenomena as they explain. All the plants of the late Paleozoic are extinct; few paleobotanists have much faith in climatic inferences from them, for there are many examples of later adaptations of plants to environments that differ widely from those in which they arose. The earth's angular momentum is great. Like a gyroscope, it strongly resists change in its axis of rotation, and astronomers do not believe it possible that the polar axis shifts in other than very small amounts.

Another good reason for doubting that the drifting of continents causes mountain folding is that, even if we grant it for the Tertiary mountain chains, we are left with no satisfactory mechanism for all the far older ones. Though the theory is a brilliant tour-de-force, its support does not seem substantial. Nevertheless, it has focused attention on one of the most difficult problems of geology: If the continents are, indeed, light masses floating on a denser substratum, how is it possible for an area once continental later to become a part of the ocean floor?

Evidence for Deep Submergence of Continental Margins

Though facies changes of many ancient formations have suggested that land areas have locally been drowned to considerable depths, the recent work of Maurice Ewing, an American geophysicist, has placed the matter beyond reasonable argument for the Atlantic coastal plain off the United States. With instruments for recording seismic waves set up in the rocks by artificial explosions, Ewing was able to trace several strata with distinctive elastic properties from their exposures on land (and in deep wells) far out to sea, in fact, to the edge of the continental shelf. In this way he determined that the rocks which lie unconformably upon the folded Appalachians form a huge embankment, in places more than

3 miles thick. The basement of this embankment has the same elastic properties as the metamorphosed rocks exposed in the Appalachian Piedmont, and it is practically certain that the unconformity separating this from the younger series has been identified for about 100 miles offshore. In other words, the eroded land surface beneath the Mesozoic sediments that we find at many places in the Piedmont, has been traced almost to the edge of the continental shelf, where it is more than 2 miles below the sea.

First, in the present context, it shows that former continental areas may become submerged to the depth of the ocean floor near the continental border; second, it suggests the probability that this continental shelf, above the crystalline floor, is a sedimentary embankment built of Mesozoic and Tertiary rocks, and not in any large degree a cut surface planed off by marine erosion, although there is no doubt that marine erosion has modified it. Although features such as these pose difficult problems, in the light of the earth's general isostatic condition, they furnish strong reason to doubt that continental drifting is necessary to bring about the present wide separation of terrestrial plant and animal fossils that we thought required former land connections like those assumed for Gondwana Land.

Another theory that may hold promise of reconciling these many diverse factors has recently come under consideration. This is the theory of mountain-building by convection currents within the body of the earth.

THE CONVECTION THEORY

Any fluid heated from below tends to lose heat in two ways: by conduction and by convection. In large masses conduction is ineffective because the amount of heat transferred in a given time decreases as the square of the thickness of the mass through which it must be transmitted. The effectiveness of convection depends on the rate of overturn of the liquid. Convection is favored by low viscosity, low conductivity (or large size, which has the same effect), and by a large increase in temperature with depth.

Now it may seem absurd to think of the subcrustal part of the earth as like a liquid subject to convective overturn, for the speed of transverse seismic waves shows that it is a rigid solid to the core boundary at 2,900-km. depth. Nevertheless, we know from the folded structures of sedimentary rocks in mountain ranges and from the contortion of many metamorphic rocks that these, though not molten, have been plastic and have flowed.

And isostasy shows that over broad regions the subcrustal material must act much like a dense liquid in which the lighter crust is floating. The viscosity of the subcrustal material has been computed from the rate of uplift of the Scandinavian areas in response to the unloading of glacial ice. It is very high, but, since the temperature increases downward, it is possible that the viscosity is less near the core, despite the tremendous pressure there. If, then, the deeper parts of the mantle contain radioactive material in amounts equal to those of meteorites, heating by radioactive disintegration will make the mantle unstable, and may start a slow convective overturn.

The theory has been analyzed by the Israeli physicist Pekeris on the idealized scheme of which, for convenience in computation, assumes two polar continents and an equatorial ocean. He found that convective currents moving more than 3 cm. per year may be present beneath such idealized continents, and that their drag on a rigid crust would be strong enough to crush the crustal rocks even if they were three times as strong as tests indicate. Other theoretical work on the problem, assuming different convective depths, yield comparable results. These computations are all based on a constant rate of convective overturn, such as would take place in liquids with no shear strength but with appropriate viscosities.

The American geophysicist Griggs, however, has pointed out that the subcrust is not a true liquid but has a certain strength, even though this is small under the high temperatures deep within the earth. Properly to evaluate these experiments and the theoretical basis on which they rest demands more background in science than can be given in this book. Nevertheless, we can see that a surprisingly large number of the phenomena connected with mountains are coordinated by this hypothesis.

Perhaps the weakest aspect of the convection theory is the formation of the geosyncline: The period assigned to slowly accelerating currents whose downward drag develops the crustal sag to be filled by sediments is much too short for most geosynclines. But, if this weakness is overlooked, we can see that overthrusting and intense deformation of the passive sedimentary cover, the dragging down of the crust to form "negative belts," and the later uplift of the mountain belt after the current stops are all nicely accounted for.

When the currents stop, the root is no longer held down and will rise isostatically. Furthermore, in being dragged down the root enters a region of higher temperature, and the highly sheared and deformed crustal rocks of which it is composed would recrystallize to metamorphic rocks and be injected by magma. The magma may be produced by liquefaction of subcrustal material which is already near the melting point and ready to "flash" into liquid in zones of fracture where pressure suddenly falls.

Some of the root may only partly remelt to form migmatites, and part may flow upward as granitic batholiths to congeal at higher levels in the mountain structure. Further, it is reasonable to believe that a softened and partly melted root may flow laterally as well as upward, thereby accounting for the common widespread uplift that affects not only the folded belt itself, but also the country alongside, the surface rocks of which are unfolded.

On such an hypothesis, the belts of negative anomaly in the East Indies and the comparable belts in the West Indies and along some of the island arcs of the Pacific, such as the Marianas and Japanese arcs, would be interpreted as due to the dragging down of the crust by existing subcrustal currents. The fact that deep-focus earthquakes seem definitely alined in planes that dip away from the ocean deeps supports the idea that such currents are actually dragging the oceanic crust beneath the land masses alongside.

The fault plane marked by the earthquake epicenters may

be thought of as a shear zone plunging down with the descending currents. Obviously, it is far too soon to accept the convection theory, at least in its present form, as a satisfactory explanation of the complexities of mountain-making. Until further knowledge is gained as to conditions deep within the earth—and such knowledge is necessarily indirect and difficult to evaluate—we cannot even be sure that the necessary driving force can be furnished.

The dearth of conclusive evidence for the other theories suggested gives no support for this one. Yet it is possible, since so many phenomena can be naturally fitted into this scheme, that it holds seeds for an ultimately satisfactory explanation of the fascinating enigma that confronts us: an earth at least two billion years old but still, beneath its passive crust, endowed with the energy literally to move mountains! Indeed, if evidence for continental displacement ever becomes so strong as to be compelling, it may be that subcrustal convection currents will offer the only mechanism by which to bring these, too, about.

Chapter 8

Plate Tectonics

A tectonic plate (also called lithospheric plate) is a massive, irregularly shaped slab of solid rock, generally composed of both continental and oceanic lithosphere. Plate size can vary greatly, from a few hundred to thousands of kilometers across; the Pacific and Antarctic Plates are among the largest. Plate thickness also varies greatly, ranging from less than 15 km for young oceanic lithosphere to about 200 km or more for ancient continental lithosphere.

How do these massive slabs of solid rock float despite their tremendous weight? The answer lies in the composition of the rocks. Continental crust is composed of granitic rocks which are made up of relatively lightweight minerals such as quartz and feldspar. By contrast, oceanic crust is composed of basaltic rocks, which are much denser and heavier. The variations in plate thickness are nature's way of partly compensating for the imbalance in the weight and density of the two types of crust. Because continental rocks are much lighter, the crust under the continents is much thicker (as much as 100 km) whereas the crust under the oceans is generally only about 5 km thick. Like icebergs, only the tips of which are visible above water, continents have deep "roots" to support their elevations.

Most of the boundaries between individual plates cannot be seen, because they are hidden beneath the oceans. Yet oceanic plate boundaries can be mapped accurately from outer space by measurements from GEOSAT satellites. Earthquake and volcanic activity is concentrated near these boundaries. Tectonic plates probably developed very early in the Earth's

4.6-billion-year history, and they have been drifting about on the surface ever since-like slow-moving bumper cars repeatedly clustering together and then separating.

Like many features on the Earth's surface, plates change over time. Those composed partly or entirely of oceanic lithosphere can sink under another plate, usually a lighter, mostly continental plate, and eventually disappear completely. This process is happening now off the coast of Oregon and Washington. The small Juan de Fuca Plate, a remnant of the formerly much larger oceanic Farallon Plate, will someday be entirely consumed as it continues to sink beneath the North American Plate.

One of the first steps in performing a scientific investigation is to understand what is already known. The investigation could be posed in a number of ways, but I have settled on asking students to find evidence for the theory of plate tectonics, as explained in their textbook.

This approach has worked well in a class of general education majors. With more class time it might be possible to lead students to the discovery that the earth's surface must actually be moving around. The seafloor age data requires it, as do the existence of earthquakes in well defined patterns. Nevertheless, the search for this evidence provides rich investigative opportunities. Furthermore, the real earth is uncooperative enough to provide many complications for deeper investigation.

The first step is to familiarize students with the theory as described in the simplistic cartoons that depict spreading centers, transform faults and subduction zones. A clear understanding of the basic theory is essential if students are to successfully find data in support of it. The "Plate Tectonics Lecture" does an excellent job of animating these cartoons to aid in visualization.

In some instances, clay models might also prove useful. The theory of plate tectonics states that the earth's surface has a number of rigid, brittle segments that move around. The outlines of these plates are easy to see on a world map of earthquakes. Earthquakes occur where brittle pieces of the earth are slipping past one another. When you break a dish, a "crack" occurs and there is a separation of two or more pieces. The noise is analogous to the earthquake and the crack is where relative motion between two pieces takes place. The plate is called of "Lithosphere". Within the lithosphere is the crust and upper portion of the mantle. There are three kinds of boundaries between plates.

DIVERGENT BOUNDARIES

These are called "spreading centers" and often occur at "mid-ocean ridges". Two plates are separating, or diverging. Location A is the location of the spreading centre in the middle of the Atlantic Ocean. Notice that the spreading centre (A) is higher in elevation (shallower water).

Even before plate tectonics, this long linear feature was described as a "midocean ridge". This is because the lithosphere is hottest at a spreading centre, so thermal expansion makes its elevation higher. Hot mantle rocks rise into the space left by the separating plates and form the new oceanic crust. The Mid-Atlantic ridge has many large volcanoes on its flanks. The region is characterized by few earthquakes, most of which are on short "transform" segments.

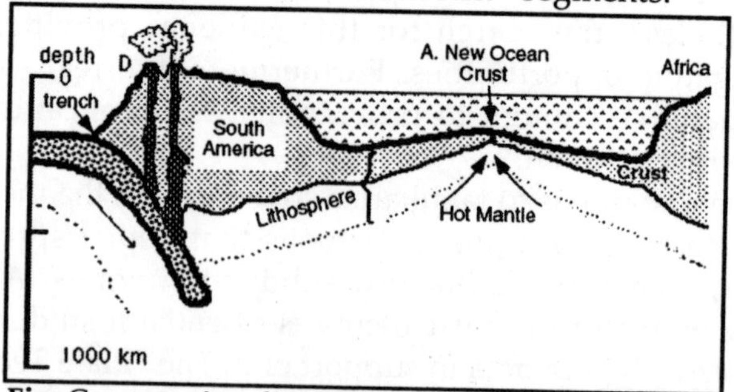

Fig. Cross section diagram of the Eastern Pacific and Atlantic Ocean basins.

TRANSFORM BOUNDARIES

Transform boundaries are where the plates slide past one another. The San Andreas Fault is a transform boundary between the Pacific and North American Plates. Transform boundaries are characterized by many earthquakes at relatively shallow depths (less that 50 km deep).

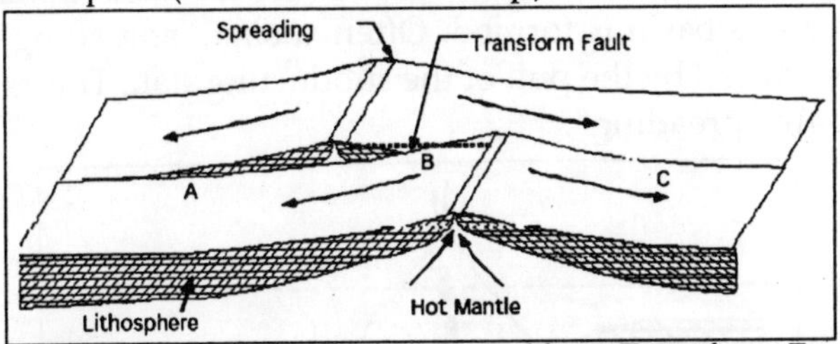

Fig. Offset Spreading Centers, Joined by a Transform Fault.

The location between the two spreading centers which is the transform fault. This is the only region where there is relative sliding motion. Each of the offset spreading centers is at a higher elevation, so when the offset spreading centers are joined by a transform fault, there are elevation differences at the boundary.. Boundary segments marked by A and C do not have relative motion across the "fault" (dotted line segment). Yet the original elevation differences caused by the offset are still present. These segments are called "Fracture Zones", and can be observed in the Map elevation data in the eastern Pacific Ocean, where they may persist for thousands of kilometers.

In map view, the representation of offset spreading centers, joined by a transform fault. The transform fault is indicated by the B, and A and C are the fracture zones, which carry the scar of the old transform fault along as the plate spreads away from the spreading centre.

Convergent Boundaries

A boundary where plates are coming together, or "converging". There are a number of important variations to converging boundaries. Collision between an oceanic plate and a continent. The collision between India and Eurasia is an example of a plate convergent boundary where two continents collide. The continents are both too buoyant to sink, so their collision causes a thickening of the lithosphere, which is manifested in the Himalayan Mountains and Tibetan Plateau. When an oceanic plate collides and subducts beneath another oceanic plate, an island arc forms. The islands are caused by the volcanoes created by the subduction process. Behind the island arc, a basin is formed. Often, a small spreading centre occurs, caused by the pull of the subducting slab. This is called "back-arc spreading".

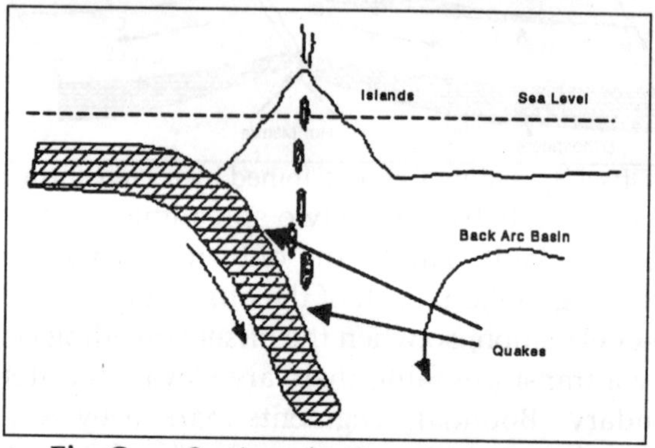

Fig. Cross Section of a Convergent Margin between two Oceanic Plates.

There are many "real world" examples of the simple structures described above. You will be able to find these using the map software. However, there are also many complications, which are beyond the scope of this treatment. One interesting complication occurs at the northern end of the San Andreas Fault. The western margin of North America is a combination of subduction zone and transform fault. North of Mendocino, beneath Northern California, Oregon, Washington, and southern Canada, there is a very slow subduction zone. This subduction zone is responsible for the active volcanoes: Mt. St. Helens, Mt. Ranier, Mt. Hood, Mt. Adams, and others. South of Mendocino is the San Andreas Fault, which veers offshore to the Gorda Rise, which is a spreading centre.

HISTORICAL PERSPECTIVE

In geologic terms, a *plate* is a large, rigid slab of solid rock. The word *tectonics* comes from the Greek root "to build." Putting these two words together, we get the term *plate tectonics,* which refers to how the Earth's surface is built of plates. The *theory of plate tectonics* states that the Earth's outermost layer is fragmented into a dozen or more large and small plates that are moving relative to one another as they ride atop hotter, more mobile material. Before the advent of plate tectonics, however, some people already believed that the present-day continents were the fragmented pieces of preexisting larger landmasses ("supercontinents"). The diagrams below show the break-up of the supercontinent *Pangaea* (meaning "all lands" in Greek), which figured prominently in the *theory of continental drift* the forerunner to the theory of plate tectonics.

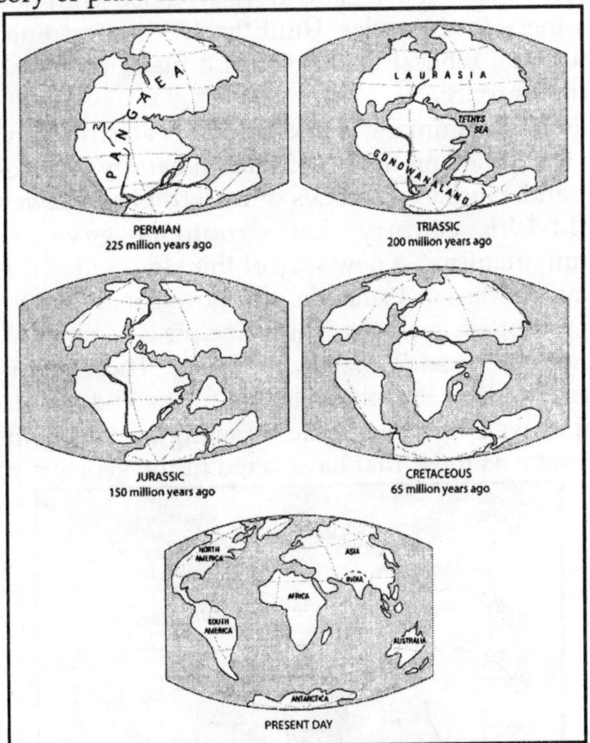

Fig. The Supercontinent Pangaea began to break up about 225-200 million years ago.

Plate tectonics is a relatively new scientific concept, introduced some 30 years ago, but it has revolutionized our understanding of the dynamic planet upon which we live. The theory has unified the study of the Earth by drawing together many branches of the earth sciences, from *paleontology* (the study of fossils) to *seismology*. It has provided explanations to questions that scientists had speculated upon for centuries — such as why earthquakes and volcanic eruptions occur in very specific areas around the world, and how and why great mountain ranges like the Alps and Himalayas formed.

Why is the Earth so restless? What causes the ground to shake violently, volcanoes to erupt with explosive force, and great mountain ranges to rise to incredible heights? Scientists, philosophers, and theologians have wrestled with questions such as these for centuries. Until the 1700s, most Europeans thought that a Biblical Flood played a major role in shaping the Earth's surface.

This way of thinking was known as "catastrophism," and *geology* was based on the belief that all earthly changes were sudden and caused by a series of catastrophes. However, by the mid-19th century, catastrophism gave way to "uniformitarianism," a new way of thinking centered around the "Uniformitarian Principle" proposed in 1785 by James Hutton, a Scottish geologist. This principle is commonly stated as follows: *The present is the key to the past.* Those holding this viewpoint assume that the geologic forces and processes — gradual as well as catastrophic — acting on the Earth today are the same as those that have acted in the geologic past.

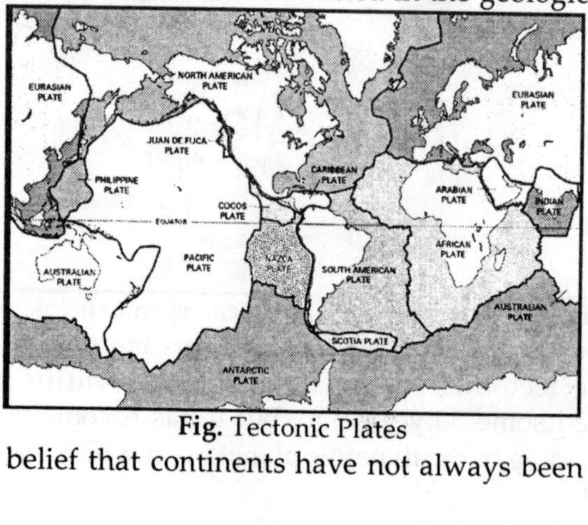

Fig. Tectonic Plates

The belief that continents have not always been fixed in

their present positions was suspected long before the 20th century; this notion was first suggested as early as 1596 by the Dutch map maker Abraham Ortelius in his work *Thesaurus Geographicus*. Ortelius suggested that the Americas were "torn away from Europe and Africa... by earthquakes and floods" and went on to say: "The vestiges of the rupture reveal themselves, if someone brings forward a map of the world and considers carefully the coasts of the three [continents]." Ortelius' idea surfaced again in the 19th century.

However, it was not until 1912 that the idea of moving continents was seriously considered as a full-blown scientific theory — called *Continental Drift* — introduced in two articles published by a 32-year-old German meteorologist named Alfred Lothar Wegener. He contended that, around 200 million years ago, the supercontinent Pangaea began to split apart. Alexander Du Toit, Professor of Geology at Johannesburg University and one of Wegener's staunchest supporters, proposed that Pangaea first broke into two large continental landmasses, *Laurasia* in the northern hemisphere and *Gondwanaland* in the southern hemisphere. Laurasia and Gondwanaland then continued to break apart into the various smaller continents that exist today.

Wegener's theory was based in part on what appeared to him to be the remarkable fit of the South American and African continents, first noted by Abraham Ortelius three centuries earlier. Wegener was also intrigued by the occurrences of unusual geologic structures and of plant and animal fossils found on the matching coastlines of South America and Africa, which are now widely separated by the Atlantic Ocean. He reasoned that it was physically impossible for most of these organisms to have swum or have been transported across the vast oceans. To him, the presence of identical fossil species along the coastal parts of Africa and South America was the most compelling evidence that the two continents were once joined.

In Wegener's mind, the drifting of continents after the break-up of Pangaea explained not only the matching fossil occurrences but also the evidence of dramatic climate changes on some continents. For example, the discovery of fossils of tropical plants (in the form of coal deposits) in Antarctica led to the conclusion that this frozen land previously must have been situated closer to the equator, in a more temperate climate where lush, swampy vegetation could grow. Other mismatches of geology and climate included distinctive fossil ferns discovered in now-polar regions, and the occurrence of glacial deposits in present-day arid Africa, such as the Vaal River valley of South Africa.

The *theory of continental drift* would become the spark that ignited a new way of viewing the Earth. But at the time Wegener introduced his theory, the scientific community firmly believed the continents and oceans to be permanent features on the Earth's surface. Not surprisingly, his proposal was not well received, even though it seemed to agree with the scientific information available at the time.

A fatal weakness in Wegener's theory was that it could not satisfactorily answer the most fundamental question raised by his critics: What kind of forces could be strong enough to move such large masses of solid rock over such great distances? Wegener suggested that the continents simply plowed through the ocean floor, but Harold Jeffreys, a noted English geophysicist, argued correctly that it was physically impossible for a large mass of solid rock to plow through the ocean floor without breaking up.

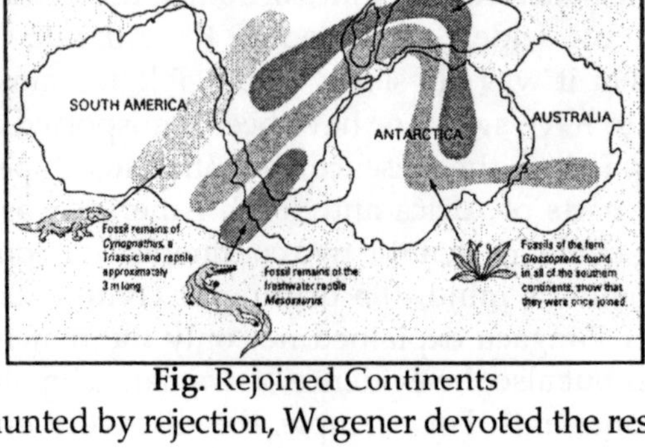

Fig. Rejoined Continents

Undaunted by rejection, Wegener devoted the rest of his life to doggedly pursuing additional evidence to defend his theory. He froze to death in 1930 during an expedition crossing the Greenland ice cap, but the controversy he spawned raged on.

However, after his death, new evidence from ocean floor exploration and other studies rekindled interest in Wegener's theory, ultimately leading to the development of the *theory of plate tectonics.* Plate tectonics has proven to be as important to the earth sciences as the discovery of the structure of the atom was to physics and chemistry and the theory of evolution was to the life sciences. Even though the theory of plate tectonics is now widely accepted by the scientific community, aspects of the theory are still being debated today. Ironically, one of the chief outstanding questions is the one Wegener failed to resolve: What is the nature of the forces propelling the plates? Scientists also debate how plate tectonics may have operated earlier in the Earth's history and whether similar processes operate, or have ever operated, on other planets in our solar system.

INSIDE THE EARTH

The size of the Earth – about 12,750 kilometers (km) in diameter-was known by the ancient Greeks, but it was not until the turn of the 20th century that scientists determined that our planet is made up of three main layers: *crust, mantle,* and *core.* This layered structure can be compared to that of a boiled egg.

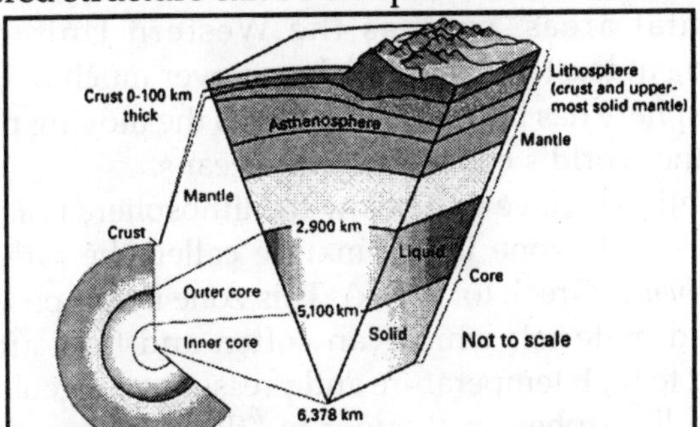

Fig. Internal Structure of the Earth.

The *crust,* the outermost layer, is rigid and very thin compared with the other two. Beneath the oceans, the crust varies little in thickness, generally extending only to about 5 km. The thickness of the crust beneath continents is much more variable but averages about 30 km; under large mountain ranges, such as the Alps or the Sierra Nevada, however, the base of the crust can be as deep as 100 km. Like the shell of an egg, the Earth's crust is brittle and can break.

Below the crust is the *mantle,* a dense, hot layer of semi-solid rock approximately 2,900 km thick. The mantle, which contains more iron, magnesium, and calcium than the crust, is hotter and denser because temperature and pressure inside the Earth increase with depth. As a comparison, the mantle might be thought of as the white of a boiled egg. At the centre of the Earth lies the *core,* which is nearly twice as dense as the mantle because its composition is metallic (iron-nickel alloy) rather than stony. Unlike the yolk of an egg, however, the Earth's core is actually made up of two distinct parts: a 2,200 km-thick *liquid* outer core and a 1,250 km-thick *solid* inner core. As the Earth rotates, the liquid outer core spins, creating the Earth's magnetic field.

Not surprisingly, the Earth's internal structure influences plate tectonics. The upper part of the mantle is cooler and more rigid than the deep mantle; in many ways, it behaves like the overlying crust. Together they form a rigid layer of rock called the *lithosphere* (from *lithos,* Greek for stone). The lithosphere tends to be thinnest under the oceans and in volcanically active continental areas, such as the Western United States. Averaging at least 80 km in thickness over much of the Earth, the lithosphere has been broken up into the moving plates that contain the world's continents and oceans.

Scientists believe that below the lithosphere is a relatively narrow, mobile zone in the mantle called the *asthenosphere* (from *asthenes,* Greek for weak). This zone is composed of hot, semi-solid material, which can soften and flow after being subjected to high temperature and pressure over geologic time. The rigid lithosphere is thought to "float" or move about on the slowly flowing asthenosphere.

DEVELOPING THE THEORY

Continental drift was hotly debated off and on for decades

following Wegener's death before it was largely dismissed as being eccentric, preposterous, and improbable. However, beginning in the 1950s, a wealth of new evidence emerged to revive the debate about Wegener's provocative ideas and their implications. In particular, four major scientific developments spurred the formulation of the plate-tectonics theory:

- Demonstration of the ruggedness and youth of the ocean floor;
- Confirmation of repeated reversals of the Earth magnetic field in the geologic past;
- Emergence of the seafloor-spreading hypothesis and associated recycling of oceanic crust; and
- Precise documentation that the world's earthquake and volcanic activity is concentrated along oceanic trenches and submarine mountain ranges.

Ocean Floor Mapping

About two thirds of the Earth's surface lies beneath the oceans. Before the 19th century, the depths of the open ocean were largely a matter of speculation, and most people thought that the ocean floor was relatively flat and featureless. However, as early as the 16th century, a few intrepid navigators, by taking soundings with hand lines, found that the open ocean can differ considerably in depth, showing that the ocean floor was not as flat as generally believed. Oceanic exploration during the next centuries dramatically improved our knowledge of the ocean floor. We now know that most of the geologic processes occurring on land are linked, directly or indirectly, to the dynamics of the ocean floor.

"Modern" measurements of ocean depths greatly increased in the 19th century, when deep-sea line soundings (*bathymetric* surveys) were routinely made in the Atlantic and Caribbean. In 1855, a bathymetric chart published by U.S. Navy Lieutenant Matthew Maury revealed the first evidence of underwater mountains in the central Atlantic (which he called "Middle Ground"). This was later confirmed by survey ships laying the trans-Atlantic telegraph cable. Our picture of the ocean floor greatly sharpened after World War I, when echo-sounding devices — primitive sonar systems — began to measure ocean depth by recording the time it took for a sound signal (commonly an electrically generated "ping") from the ship to bounce off the ocean floor and return. Time graphs of the returned signals revealed that the ocean floor was much more rugged than previously thought. Such echo-sounding measurements clearly demonstrated the continuity and roughness of the submarine mountain chain in the central Atlantic (later called the *Mid-Atlantic Ridge*) suggested by the earlier bathymetric measurements.

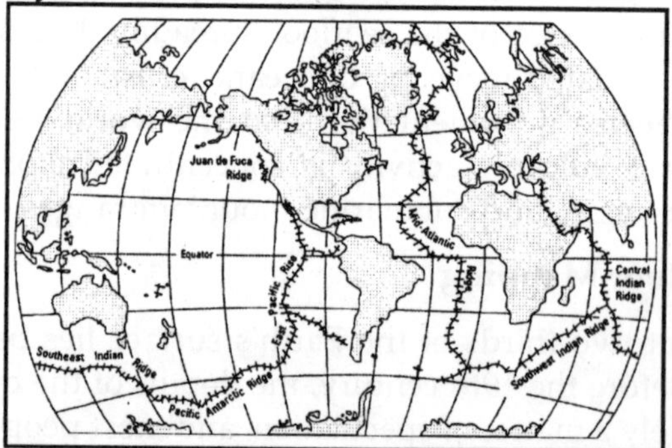

Fig. The Mid-ocean Ridge (shown in red) Winds its way between the Continents much like the Seam on a Baseball

In 1947, seismologists on the U.S. research ship *Atlantis* found that the sediment layer on the floor of the Atlantic was much thinner than originally thought. Scientists had previously believed that the oceans have existed for at least 4 billion years, so therefore the sediment layer should have been very thick. Why then was there so little accumulation of sedimentary rock and debris on the ocean floor? The answer to this question, which came after further exploration, would prove to be vital to advancing the concept of plate tectonics.

In the 1950s, oceanic exploration greatly expanded. Data gathered by oceanographic surveys conducted by many nations led to the discovery that a great mountain range on the ocean floor virtually encircled the Earth. Called the *global mid-ocean ridge,* this immense submarine mountain chain more than 50,000 kilometers (km) long and, in places, more than 800 km across — zig-zags between the continents, winding its way around the globe like the seam on a baseball. Rising an average of about 4,500 meters(m) above the sea floor, the mid-ocean ridge overshadows all the mountains in the United States except for Mount McKinley (Denali) in Alaska (6,194 m). Though hidden beneath the ocean surface, the global mid-ocean ridge system is the most prominent topographic feature on the surface of our planet.

Magnetic Striping and Polar Reversals

Beginning in the 1950s, scientists, using magnetic instruments *(magnetometers)* adapted from airborne devices developed during World War II to detect submarines, began recognizing odd magnetic variations across the ocean floor. This finding, though unexpected, was not entirely surprising because it was known that *basalt* the iron-rich, volcanic rock making up the ocean floor contains a strongly magnetic mineral *(magnetite)* and can locally distort compass readings. This distortion was recognized by Icelandic mariners as early as the late 18th century. More important, because the presence of magnetite gives the basalt measurable magnetic properties, these newly discovered magnetic variations provided another means to study the deep ocean floor.

Early in the 20th century, *paleomagnetists* (those who study the Earth's ancient magnetic field) such as Bernard Brunhes in France (in 1906) and Motonari Matuyama in Japan (in the 1920s) — recognized that rocks generally belong to two groups according to their magnetic properties. One group has so-called *normal polarity,* characterized by the magnetic minerals in the rock having the same polarity as that of the Earth's present magnetic field.

This would result in the north end of the rock's "compass needle" pointing toward magnetic north. The other group, however, has *reversed polarity,* indicated by a polarity alignment opposite to that of the Earth's present magnetic field. In this case, the north end of the rock's compass needle would point south. How could this be? This answer lies in the magnetite in volcanic rock. Grains of magnetite — behaving like little magnets — can align themselves with the orientation of the Earth's magnetic field. When *magma* (molten rock containing minerals and gases) cools to form solid volcanic rock, the alignment of the magnetite grains is "locked in," recording the Earth's magnetic orientation or polarity (normal or reversed) at the time of cooling.

As more and more of the seafloor was mapped during the 1950s, the magnetic variations turned out not to be random or isolated occurrences, but instead revealed recognizable patterns. When these magnetic patterns were mapped over a wide region, the ocean floor showed a zebra-like pattern. Alternating stripes of magnetically different rock were laid out in rows on either side of the mid-ocean ridge: one stripe with normal polarity and the adjoining stripe with reversed polarity. The overall pattern, defined by these alternating bands of normally and reversely polarized rock, became known as magnetic striping.

SEAFLOOR SPREADING AND RECYCLING OF OCEANIC CRUST

The discovery of magnetic striping naturally prompted more questions: How does the magnetic striping pattern form? And why are the stripes symmetrical around the crests of the mid-ocean ridges? These questions could not be answered without also knowing the significance of these ridges. In 1961, scientists began to theorize that mid-ocean ridges mark structurally weak zones where the ocean floor was being ripped in two lengthwise along the ridge crest. New magma from deep within the Earth rises easily through these weak zones and eventually erupts along the crest of the ridges to create new oceanic crust. This process, later called *seafloor spreading,* operating over many millions of years has built the 50,000 km-long system of mid-ocean ridges. This hypothesis was supported by several lines of evidence:

- At or near the crest of the ridge, the rocks are very young, and they become progressively older away from the ridge crest
- The youngest rocks at the ridge crest always have present-day (normal) polarity; and

- Stripes of rock parallel to the ridge crest alternated in magnetic polarity (normal-reversed-normal, etc.), suggesting that the Earth's magnetic field has flip-flopped many times. By explaining both the zebralike magnetic striping and the construction of the mid-ocean ridge system, the seafloor spreading hypothesis quickly gained converts and represented another major advance in the development of the plate-tectonics theory. Furthermore, the oceanic crust now came to be appreciated as a natural "tape recording" of the history of the reversals in the Earth's magnetic field.

Additional evidence of seafloor spreading came from an unexpected source: petroleum exploration. In the years following World War II, continental oil reserves were being depleted rapidly and the search for offshore oil was on. To conduct offshore exploration, oil companies built ships equipped with a special drilling rig and the capacity to carry many kilometers of drill pipe.

This basic idea later was adapted in constructing a research vessel, named the *Glomar Challenger*, designed specifically for marine geology studies, including the collection of drill-core samples from the deep ocean floor. In 1968, the vessel embarked on a year-long scientific expedition, criss-crossing the Mid-Atlantic Ridge between South America and Africa and drilling core samples at specific locations. When the ages of the samples were determined by paleontologic and isotopic dating studies, they provided the clinching evidence that proved the seafloor spreading hypothesis.

A profound consequence of seafloor spreading is that new crust was, and is now, being continually created along the oceanic ridges. This idea found great favour with some scientists who claimed that the shifting of the continents can be simply explained by a large increase in size of the Earth since its formation. However, this so-called "expanding Earth" hypothesis was unsatisfactory because its supporters could offer no convincing geologic mechanism to produce such a huge, sudden expansion. Most geologists believe that the Earth has changed little, if at all, in size since its formation 4.6 billion years ago, raising a key question: how can new crust be continuously added along the oceanic ridges without increasing the size of the Earth?

This question particularly intrigued Harry H. Hess, a Princeton University geologist and a Naval Reserve Rear Admiral, and Robert S. Dietz, a scientist with the U.S. Coast and Geodetic Survey who first coined the term *seafloor spreading*. Dietz and Hess were among the small handful who really understood the broad implications of sea floor spreading. If the Earth's crust was expanding along the oceanic ridges, Hess reasoned, it must be shrinking elsewhere. He suggested that new oceanic crust continuously spread away from the ridges in a conveyor belt-like motion.

Many millions of years later, the oceanic crust eventually descends into the oceanic *trenches* – very deep, narrow canyons along the rim of the Pacific Ocean basin. According to Hess, the Atlantic Ocean was expanding while the Pacific Ocean was shrinking. As old oceanic crust was consumed in the trenches, new magma rose and erupted along the spreading ridges to form new crust. In effect, the ocean basins were perpetually being "recycled," with the creation of new crust and the destruction of old oceanic lithosphere occurring simultaneously. Thus, Hess' ideas neatly explained why the Earth does not get bigger with sea floor spreading, why there is so little sediment accumulation on the ocean floor, and why oceanic rocks are much younger than continental rocks.

CONCENTRATION OF EARTHQUAKES

During the 20th century, improvements in seismic instrumentation and greater use of earthquake-recording instruments *(seismographs)* worldwide enabled scientists to learn that earthquakes tend to be concentrated in certain areas, most notably along the oceanic trenches and spreading ridges. By the late 1920s, seismologists were beginning to identify several prominent earthquake zones parallel to the trenches that typically were inclined 40-60° from the horizontal and extended several hundred kilometers into the Earth. These zones later became known as *Wadati-Benioff zones,* or simply

Benioff zones, in honor of the seismologists who first recognized them, Kiyoo Wadati of Japan and Hugo Benioff of the United States. The study of global seismicity greatly advanced in the 1960s with the establishment of the Worldwide Standardized Seismograph Network monitor the compliance of the 1963 treaty banning above-ground testing of nuclear weapons. The much-improved data from the WWSSN instruments allowed seismologists to map precisely the zones of earthquake concentration worldwide.

But what was the significance of the connection between earthquakes and oceanic trenches and ridges? The recognition of such a connection helped confirm the seafloor-spreading hypothesis by pin-pointing the zones where Hess had predicted oceanic crust is being generated (along the ridges) and the zones where oceanic lithosphere sinks back into the mantle (beneath the trenches).

UNDERSTANDING PLATE MOTION

Scientists now have a fairly good understanding of how the plates move and how such movements relate to earthquake activity. Most movement occurs along narrow zones between plates where the results of plate-tectonic forces are most evident.

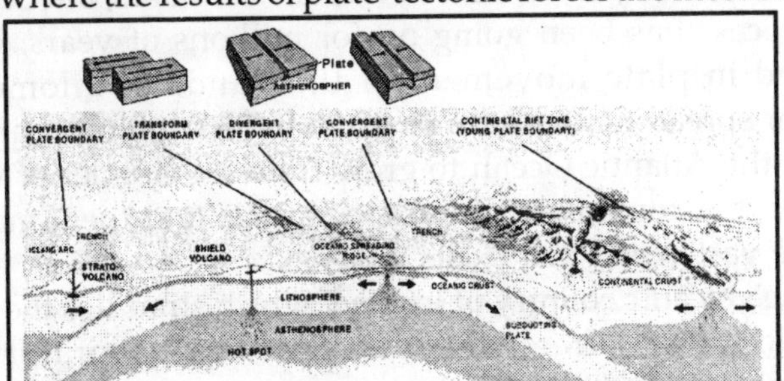

Fig. The Main Types of Plate Boundaries

There are four types of plate boundaries:

- Divergent boundaries where new crust is generated as the plates pull away from each other.
- Convergent boundaries where crust is destroyed as one plate dives under another.

- Transform boundaries where crust is neither produced nor destroyed as the plates slide horizontally past each other.
- Plate boundary zones broad belts in which boundaries are not well defined and the effects of plate interaction are unclear.

Divergent boundaries

Divergent boundaries occur along spreading centers where plates are moving apart and new crust is created by magma pushing up from the mantle. Picture two giant conveyor belts, facing each other but slowly moving in opposite directions as they transport newly formed oceanic crust away from the ridge crest.

Perhaps the best known of the divergent boundaries is the Mid-Atlantic Ridge. This submerged mountain range, which extends from the Arctic Ocean to beyond the southern tip of Africa, is but one segment of the global mid-ocean ridge system that encircles the Earth. The rate of spreading along the Mid-Atlantic Ridge averages about 2.5 centimeters per year (cm/yr), or 25 km in a million years.

This rate may seem slow by human standards, but because this process has been going on for millions of years, it has resulted in plate movement of thousands of kilometers. Seafloor spreading over the past 100 to 200 million years has caused the Atlantic Ocean to grow from a tiny inlet of water between the continents of Europe, Africa, and the Americas into the vast ocean that exists today.

The volcanic country of Iceland, which straddles the Mid-Atlantic Ridge, offers scientists a natural laboratory for studying on land the processes also occurring along the submerged parts of a spreading ridge. Iceland is splitting along the spreading centre between the North American and Eurasian Plates, as North America moves westward relative to Eurasia.

The consequences of plate movement are easy to see around Krafla Volcano, in the northeastern part of Iceland. Here, existing ground cracks have widened and new ones appear every few months. From 1975 to 1984, numerous episodes of *rifting* (surface cracking) took place along the Krafla fissure zone. Some of these rifting events were accompanied by volcanic activity; the ground would gradually rise 1-2 m before abruptly dropping, signalling an impending eruption.

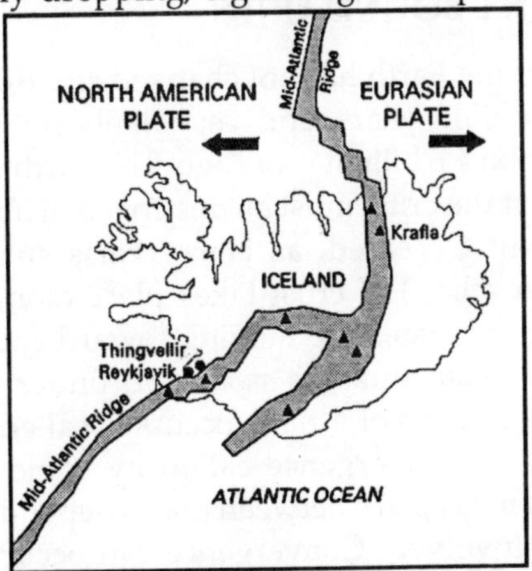

Fig. Mid-Atlantic Ridge splitting Iceland and Separating the North American and Eurasian Plates.

Between 1975 and 1984, the displacements caused by rifting totalled about 7 m. In East Africa, spreading processes have already torn Saudi Arabia away from the rest of the African continent, forming the Red Sea. The actively splitting African Plate and the Arabian Plate meet in what geologists call a *triple junction*, where the Red Sea meets the Gulf of Aden.

A new spreading centre may be developing under Africa along the East African Rift Zone. When the continental crust stretches beyond its limits, tension cracks begin to appear on the Earth's surface. Magma rises and squeezes through the widening cracks, sometimes to erupt and form volcanoes. The rising magma, whether or not it erupts, puts more pressure on the crust to produce additional fractures and, ultimately, the rift zone.

East Africa may be the site of the Earth's next major ocean. Plate interactions in the region provide scientists an opportunity to study first hand how the Atlantic may have begun to form about 200 million years ago. Geologists believe that, if spreading continues, the three plates that meet at the edge of the present-day African continent will separate completely, allowing the Indian Ocean to flood the area and making the easternmost corner of Africa a large island.

CONVERGENT BOUNDARIES

The size of the Earth has not changed significantly during the past 600 million years, and very likely not since shortly after its formation 4.6 billion years ago. The Earth's unchanging size implies that the crust must be destroyed at about the same rate as it is being created, as Harry Hess surmised. Such destruction (recycling) of crust takes place along convergent boundaries where plates are moving toward each other, and sometimes one plate sinks (is *subducted*) under another. The location where sinking of a plate occurs is called a *subduction zone.* The type of convergence called by some a very slow "collision" that takes place between plates depends on the kind of lithosphere involved. Convergence can occur between an oceanic and a largely continental plate, or between two largely oceanic plates, or between two largely continental plates.

Oceanic-continental Convergence

If by magic we could pull a plug and drain the Pacific Ocean, we would see a most amazing sight — a number of long narrow, curving *trenches* thousands of kilometers long and 8 to 10 km deep cutting into the ocean floor. Trenches are the deepest parts of the ocean floor and are created by subduction.

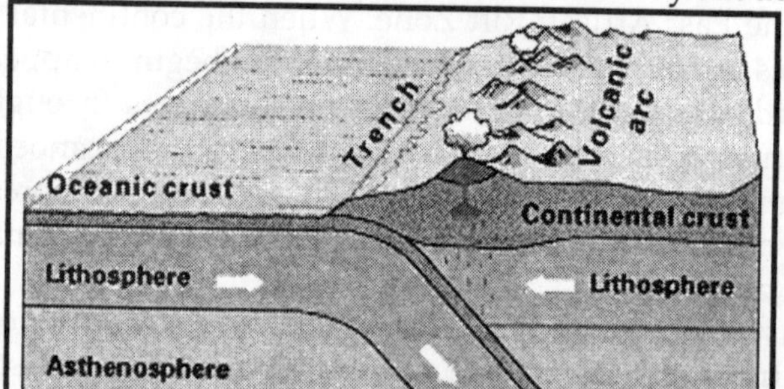

Fig. Oceanic-continental Covergence

Off the coast of South America along the Peru-Chile trench, the oceanic Nazca Plate is pushing into and being subducted under the continental part of the South American Plate. In turn, the overriding South American Plate is being lifted up, creating the towering Andes mountains, the backbone of the continent. Strong, destructive earthquakes and the rapid uplift of mountain ranges are common in this region. Even though the Nazca Plate as a whole is sinking smoothly and continuously into the trench, the deepest part of the subducting plate breaks into smaller pieces that become locked in place for long periods of time before suddenly moving to generate large earthquakes. Such earthquakes are often accompanied by uplift of the land by as much as a few meters.

On 9 June 1994, a magnitude-8.3 earthquake struck about 320 km northeast of La Paz, Bolivia, at a depth of 636 km. This earthquake, within the subduction zone between the Nazca Plate and the South American Plate, was one of deepest and largest subduction earthquakes recorded in South America. Fortunately, even though this powerful earthquake was felt as far away as Minnesota and Toronto, Canada, it caused no major damage because of its great depth.

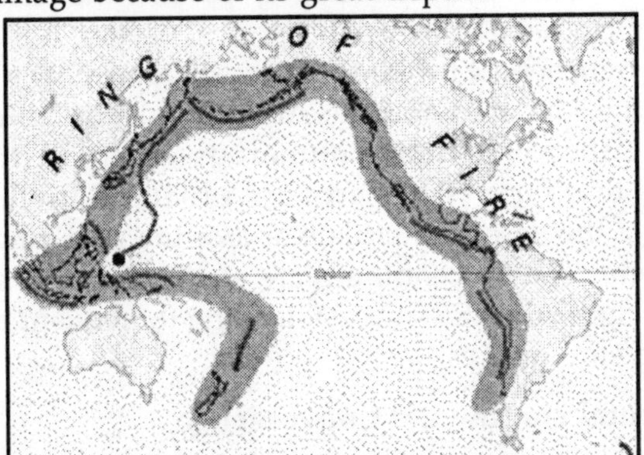

Fig. Ring of Fire

Oceanic-continental convergence also sustains many of the Earth's active volcanoes, such as those in the Andes and the Cascade Range in the Pacific Northwest. The eruptive activity is clearly associated with subduction, but scientists vigorously debate the possible sources of magma: Is magma generated by the partial melting of the subducted oceanic slab, or the overlying continental lithosphere, or both?

Oceanic-oceanic Convergence

As with oceanic-continental convergence, when two oceanic plates converge, one is usually subducted under the other, and in the process a trench is formed. The Marianas Trench for example, marks where the fast-moving Pacific Plate converges against the slower moving Philippine Plate. The Challenger Deep, at the southern end of the Marianas Trench, plunges deeper into the Earth's interior (nearly 11,000 m) than Mount Everest, the world's tallest mountain, rises above sea level (about 8,854 m).

Subduction processes in oceanic-oceanic plate convergence also result in the formation of volcanoes. Over millions of years, the erupted lava and volcanic debris pile up on the ocean floor until a submarine volcano rises above sea level to form an island volcano. Such volcanoes are typically strung out in chains called *island arcs*. As the name implies, volcanic island arcs, which closely parallel the trenches, are generally curved.

The trenches are the key to understanding how island arcs such as the Marianas and the Aleutian Islands have formed and why they experience numerous strong earthquakes. Magmas that form island arcs are produced by the partial melting of the descending plate and/or the overlying oceanic lithosphere. The descending plate also provides a source of stress as the two plates interact, leading to frequent moderate to strong earthquakes.

Continental-continental Convergence

The Himalayan mountain range dramatically demonstrates one of the most visible and spectacular consequences of plate tectonics. When two continents meet head-on, neither is subducted because the continental rocks are relatively light and, like two colliding icebergs, resist downward motion. Instead, the crust tends to buckle and be pushed upward or sideways. The collision of India into Asia 50 million years ago caused the Eurasian Plate to crumple up and override the Indian Plate.

After the collision, the slow continuous convergence of the two plates over millions of years pushed up the Himalayas and the Tibetan Plateau to their present heights. Most of this growth occurred during the past 10 million years. The Himalayas, towering as high as 8,854 m above sea level, form the highest continental mountains in the world. Moreover, the neighboring Tibetan Plateau, at an average elevation of about 4,600 m, is higher than all the peaks in the Alps except for Mont Blanc and Monte Rosa, and is well above the summits of most mountains in the United States.

TRANSFORM BOUNDARIES

The zone between two plates sliding horizontally past one another is called a *transform-fault boundary,* or simply a *transform boundary.* The concept of transform faults originated with Canadian geophysicist J. Tuzo Wilson, who proposed that these large faults or *fracture zones* connect two spreading centers (divergent plate boundaries) or, less commonly, trenches (convergent plate boundaries). Most transform faults are found on the ocean floor. They commonly offset the active spreading ridges, producing zig-zag plate margins, and are generally defined by shallow earthquakes. However, a few occur on land, for example the San Andreas fault zone in California.

This transform fault connects the East Pacific Rise, a divergent boundary to the south, with the South Gorda — Juan de Fuca Explorer Ridge, another divergent boundary to the north. The San Andreas fault zone, which is about 1,300 km long and in places tens of kilometers wide, slices through two thirds of the length of California. Along it, the Pacific Plate has been grinding horizontally past the North American Plate for 10 million years, at an average rate of about 5 cm/yr. Land on the west side of the fault zone (on the Pacific Plate) is moving in a northwesterly direction relative to the land on the east side of the fault zone.

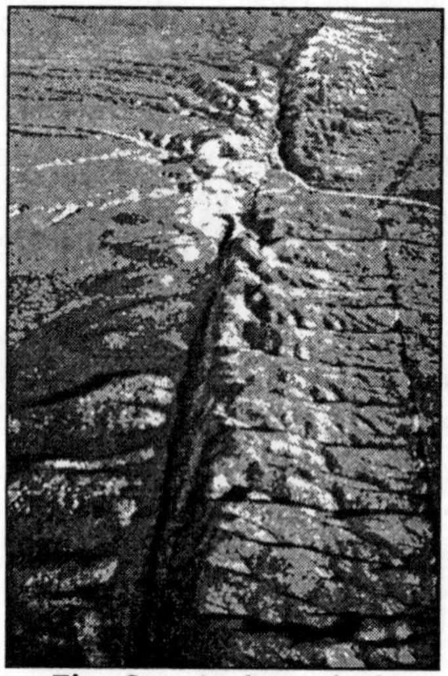

Fig. San Andreas fault

Oceanic fracture zones are ocean-floor valleys that horizontally offset spreading ridges; some of these zones are hundreds to thousands of kilometers long and as much as 8 km deep. Examples of these large scars include the Clarion, Molokai, and Pioneer fracture zones in the Northeast Pacific off the coast of California and Mexico. These zones are presently inactive, but the offsets of the patterns of magnetic striping provide evidence of their previous transform-fault activity.

PLATE-BOUNDARY ZONES

Not all plate boundaries are as simple as the main types. In some regions, the boundaries are not well defined because the plate-movement deformation occurring there extends over a broad belt (called a *plate-boundary zone*).

One of these zones marks the Mediterranean-Alpine region between the Eurasian and African Plates, within which several smaller fragments of plates *(microplates)* have been recognized. Because plate-boundary zones involve at least two large plates and one or more microplates caught up between them, they tend to have complicated geological structures and earthquake patterns.

Rates of Motion

We can measure how fast tectonic plates are moving today, but how do scientists know what the rates of plate movement have been over geologic time? The oceans hold one of the key pieces to the puzzle. Because the ocean-floor magnetic striping records the flip-flops in the Earth's magnetic field, scientists, knowing the approximate duration of the reversal, can calculate the average rate of plate movement during a given time span.

These average rates of plate separations can range widely. The Arctic Ridge has the slowest rate (less than 2.5 cm/yr), and the East Pacific Rise near Easter Island, in the South Pacific about 3,400 km west of Chile, has the fastest rate (more than 15 cm/yr).

Evidence of past rates of plate movement also can be obtained from geologic mapping studies. If a rock formation of known age with distinctive composition, structure, or fossils mapped on one side of a plate boundary can be matched with the same formation on the other side of the boundary, then measuring the distance that the formation has been offset can give an estimate of the average rate of plate motion. This simple but effective technique has been used to determine the rates of plate motion at divergent boundaries, for example the Mid-Atlantic Ridge, and transform boundaries, such as the San Andreas Fault.

Current plate movement can be tracked directly by means of ground-based or space-based *geodetic* measurements; *geodesy* is the science of the size and shape of the Earth. Ground-based measurements are taken with conventional but very precise ground-surveying techniques, using laser-electronic instruments. However, because plate motions are global in scale, they are best measured by satellite-based methods.

The late 1970s witnessed the rapid growth of *space geodesy,* a term applied to space-based techniques for taking precise, repeated measurements of carefully chosen points on the Earth's surface separated by hundreds to thousands of kilometers.

The three most commonly used space-geodetic techniques very long baseline interferometry (VLBI), satellite laser ranging (SLR), and the Global Positioning System (GPS) are based on technologies developed for military and aerospace research, notably radio astronomy and satellite tracking.

Among the three techniques, to date the GPS has been the most useful for studying the Earth's crustal movements. Twenty-one satellites are currently in orbit 20,000 km above the Earth as part of the NavStar system of the U.S. Department of Defence. These satellites continuously transmit radio signals back to Earth. To determine its precise position on Earth (longitude, latitude, elevation), each GPS ground site must simultaneously receive signals from at least four satellites, recording the exact time and location of each satellite when its signal was received.

By repeatedly measuring distances between specific points, geologists can determine if there has been active movement along faults or between plates. The separations between GPS sites are already being measured regularly around the Pacific basin.

By monitoring the interaction between the Pacific Plate and the surrounding, largely continental plates, scientists hope to learn more about the events building up to earthquakes and volcanic eruptions in the circum-Pacific Ring of Fire. Space-geodetic data have already confirmed that the rates and direction of plate movement, averaged over several years, compare well with rates and direction of plate movement averaged over millions of years.

HOTSPOTS: MANTLE THERMAL PLUMES

The vast majority of earthquakes and volcanic eruptions occur near plate boundaries, but there are some exceptions. For example, the Hawaiian Islands, which are entirely of volcanic origin, have formed in the middle of the Pacific Ocean more than 3,200 km from the nearest plate boundary. How do the Hawaiian Islands and other volcanoes that form in the interior of plates fit into the plate-tectonics picture?

In 1963, J. Tuzo Wilson, the Canadian geophysicist who discovered transform faults, came up with an ingenious idea that became known as the "hotspot" theory. Wilson noted that in certain locations around the world, such as Hawaii, volcanism has been active for very long periods of time. This could only happen, he reasoned, if relatively small, long-lasting, and exceptionally hot regions called *hotspots* existed below the plates that would provide localized sources of high heat energy *(thermal plumes)* to sustain volcanism.

Specifically, Wilson hypothesized that the distinctive linear shape of the Hawaiian Island-Emperor Seamounts chain resulted from the Pacific Plate moving over a deep, stationary hotspot in the mantle, located beneath the present-day position of the Island of Hawaii. Heat from this hotspot produced a persistent source of magma by partly melting the overriding Pacific Plate.

The magma, which is lighter than the surrounding solid rock, then rises through the mantle and crust to erupt onto the seafloor, forming an active seamount. Over time, countless eruptions cause the seamount to grow until it finally emerges above sea level to form an island volcano. Wilson suggested that continuing plate movement eventually carries the island beyond the hotspot, cutting it off from the magma source, and volcanism ceases.

As one island volcano becomes extinct, another develops over the hotspot, and the cycle is repeated. This process of volcano growth and death, over many millions of years, has left a long trail of volcanic islands and seamounts across the Pacific Ocean floor.

According to Wilson's hotspot theory, the volcanoes of the Hawaiian chain should get progressively older and become more eroded the farther they travel beyond the hotspot. The oldest volcanic rocks on Kauai, the northwesternmost inhabited Hawaiian island, are about 5.5 million years old and are deeply eroded. By comparison, on the "Big Island" of Hawaii — southeasternmost in the chain and presumably still positioned over the hotspot — the oldest exposed rocks are less than 0.7 million years old and new volcanic rock is continually being formed.

The possibility that the Hawaiian Islands become younger

to the southeast was suspected by the ancient Hawaiians, long before any scientific studies were done. During their voyages, sea-faring Hawaiians noticed the differences in erosion, soil formation, and vegetation and recognized that the islands to the northwest were older than those to the southeast. This idea was handed down from generation to generation in the legends of Pele, the fiery Goddess of Volcanoes.

Pele originally lived on Kauai. When her older sister Namakaokahai, the Goddess of the Sea, attacked her, Pele fled to the Island of Oahu.

When she was forced by Namakaokahai to flee again, Pele moved southeast to Maui and finally to Hawaii, where she now lives in the Halemaumau Crater at the summit of Kilauea Volcano.

The mythical flight of Pele from Kauai to Hawaii, which alludes to the eternal struggle between the growth of volcanic islands from eruptions and their later erosion by ocean waves, is consistent with geologic evidence obtained centuries later that clearly shows the islands becoming younger from northwest to southeast.

Although Hawaii is perhaps the best known hotspot, others are thought to exist beneath the oceans and continents. More than a hundred hotspots beneath the Earth's crust have been active during the past 10 million years. Most of these are located under plate interiors but some occur near diverging plate boundaries. Some are concentrated near the mid-oceanic ridge system, such as beneath Iceland, the Azores, and the Galapagos Islands.

A few hotspots are thought to exist below the North American Plate. Perhaps the best known is the hotspot presumed to exist under the continental crust in the region of Yellowstone National Park in northwestern Wyoming. Here are several *calderas* (large craters formed by the ground collapse accompanying explosive volcanism) that were produced by three gigantic eruptions during the past two million years, the most recent of which occurred about 600,000 years ago. Ash deposits from these powerful eruptions have been mapped as far away as Iowa, Missouri, Texas, and even northern Mexico.

The thermal energy of the presumed Yellowstone hotspot fuels more than 10,000 hot pools and springs, geysers (like Old Faithful), and bubbling *mudpots* (pools of boiling mud).

A large body of magma, capped by a *hydrothermal system* (a zone of pressurized steam and hot water), still exists beneath the caldera. Recent surveys demonstrate that parts of the Yellowstone region rise and fall by as much as 1 cm each year, indicating the area is still geologically restless.

However, these measurable ground movements, which most likely reflect hydrothermal pressure changes, do not necessarily signal renewed volcanic activity in the area.

Index

G

H

I

J

K

L

M

N

O

P